Photographic Guide
to Minerals of the World

PHOTOGRAPHIC GUIDE TO MINERALS OF THE WORLD

OLE JOHNSEN

OXFORD
UNIVERSITY PRESS

OXFORD
UNIVERSITY PRESS

Great Clarendon Street, Oxford OX2 6DP

Oxford University Press is a department of the University of Oxford.
It furthers the University's objective of excellence in research, scholarship,
and education by publishing worldwide in

Oxford New York
Athens Auckland Bangkok Bogotá Bombay Buenos Aires Calcutta
Cape Town Dar es Salaam Delhi Florence Hong Kong Istanbul
Karachi Kuala Lumpur Madrid Melbourne Mexico City Mumbai
Nairobi Paris São Paulo Shanghai Singapore Taipei Tokyo Toronto Warsaw
and associated companies in Berlin Ibadan

Oxford is a registered trade mark of Oxford University Press
in the UK and in certain other countries

Published in the United States and Canada by Princeton University Press, 41 William Street, Princeton,
New Jersey 08540

Originally published under the title: Mineralernes verden

British Library Cataloguing in Publication Data
Data available

Library of Congress Cataloging in Publication Data
Data available

ISBN 0 19 851568 5

10 9 8 7 6 5 4 3 2 1

Typeset by Narayana Press
Printed in Denmark
on acid-free paper by Narayana Press

Contents

Preface

It has been a challenge and an exciting task to write and illustrate this book, which aims to provide a comprehensive treatment of minerals. Of about five hundred minerals described here in some detail, some two hundred are fairly widespread, but the rest are less common. Nearly four thousand minerals are known at present. Some of these obviously had to be included in the book; others could equally obviously be left out, and a large intermediate group had to be more closely evaluated. I began by searching for objective criteria for selection, but in the end I made a personal choice. Other authors would no doubt have chosen differently. That's the way it is!

I have written about minerals and their properties as they can be seen with the naked eye or with a hand lens, presenting just a few examples of the intimate coherence between the internal structure of minerals and their external properties. Thus, topics such as crystal optics and X-ray crystallography, conventionally included in mineralogy books, have been excluded. Optical and X-ray methods are not, after all, very useful in the field or at home when working on a mineral collection.

A few words are called for on the use of mineral names in this book. Most minerals are referred to by their exact names, but for reasons of simplicity and practicality only 'root names' are used for a few: e.g. apatite, although this term in reality encompasses a number of closely related mineral species (fluorapatite, hydroxylapatite, etc.), which are usually indistinguishable by eye. For the same reasons suffixes such as the Levinson suffix, e.g. -(Y) in fergusonite -(Y), are generally omitted and are mentioned only in special cases.

Drawings or colour photographs are used to illustrate nearly all the minerals described.

The photographs in particular deserve comment. A perfect photograph of a mineral would ideally show its typical appearance, underline its particular characteristics, and illustrate its mineral association, while at the same time displaying the aesthetic qualities of the mineral. In practice, every photograph is a compromise and choices have to be made. My choice was to portray mineral specimens that are of above-average quality. This brings into prominence well-developed crystals that better display the external characteristics of the mineral, if to some extent at the expense of showing a more typical specimen.

All the photographed specimens—with the exception of five from private collections—belong to the Mineral Collection of the Geological Museum, University of Copenhagen.

In order to save space, the mineral descriptions alternate between a telegraphic style in the listing of properties and a more narrative style in the sections on mineral occurrences. Precise technical terms are generally used here rather than lengthy paraphrases. All technical terms are explained, either in the introductory sections on mineral properties or in the glossary at the end of the book. These terms will also be found in the index. Both names and symbols are given for the chemical elements in the introductory sections; in the mineral descriptions symbols alone are used. A charge is shown with a symbol only when this is relevant to the context.

It would not have been possible to write this book without a great deal of support from my institution. I should like to thank the Geological Museum, University of Copenhagen, and in particular my close colleague Ole V. Petersen for his continued interest during the preparation of this book. I am especially indebted to Professor Emeritus Harry Micheelsen for his thorough examination of a first draft; his many constructive comments have been extremely helpful. My thanks are also due to Gunnar Raade, of the Geological Museum, University of Oslo, Fred Steinar Nordrum, of the Norwegian Mining Museum, Kongsberg, Dan Holstam, of the Swedish Museum of Natural History, Stockholm, and Robert A. Gault, of the Canadian Museum of Nature, Ottawa, for checking many of the mineral localites. Regardless of the valuable assistance received from colleagues, I am naturally the only person responsible for any errors that may occur.

February 2002   OLE JOHNSEN

Note on geographical names
The spelling of geographical names in the text follows that of *The Times Atlas*.

MINERALOGY AND CRYSTALLOGRAPHY

What is a mineral?

The word *mineral* is used in several ways. In geography books we read about countries with large mineral resources; in weekly magazines about healthy food with plenty of vitamins and minerals; and in advertisements about special offers on mineral water. In the old game of 'Animal, Mineral, or Vegetable?' we begin by asking whether 'it' belongs to the animal kingdom, the mineral kingdom, or the vegetable kingdom, and thus acknowledge the traditional threefold classification of nature in which all inorganic or non-living materials are assigned to the mineral kingdom. How do these ideas fit together?—or rather, what exactly *is* a mineral? It seems natural to ask the specialists—the mineralogists—this question. They say:

> 'In general terms, a mineral is an element or chemical compound that is normally crystalline and that has been formed as a result of geological processes' (E.H. Nickel (1995), *Canadian Mineralogist* 33, p. 689).

That is brief and to the point, concentrating on the essential criteria. First, the definition tells us that a mineral is formed by a geological process, i.e. by natural means, without human interference. By definition, such processes take place on the earth—geology is the science of the earth—but they are also known from other parts of the solar system. A geological process does not take place in a laboratory, even though it can be reproduced there, and this implies that materials produced synthetically are not minerals. A diamond found in a rock is thus a mineral, whereas a diamond made in the laboratory is not; it should be called a synthetic diamond or something else. This may seem somewhat pedantic, but there are a number of good reasons for making the distinction. We need not go into further detail here.

The second requirement in the definition of a mineral is that it is normally crystalline, i.e. built up as a crystal. A crystal is a solid bounded by plane faces that are an external expression of an internal order in which the chemical constituents are located in a regular three-dimensional lattice: a *crystal lattice*. A solid is crystalline when it possesses this internal order. Plane faces are formed only under particular conditions of growth. A crystalline material is thus characterized by two things: (i) its chemical composition, i.e. the elements of which it is made up, and (ii) the manner in which these elements are ordered in a crystal lattice (including the dimensions and symmetry of the lattice). It follows from this that a crystalline material is *homogeneous* in the sense that it has the same properties, physically and chemically, throughout, and that these properties are always found in the same material, regardless of the environment in which it is formed. Quartz is quartz, regardless of its place of formation.

A few minerals, such as chalcedony, have a special crystalline structure that does not result in normal crystals with plane faces. A few other minerals are not crystalline at all, either because they were never crystalline or because their crystal lattices have been destroyed by radioactivity. Such compounds can nevertheless be accepted as minerals because they are of geological origin and because they are sufficiently homogeneous and thus possess well-defined chemical and physical properties.

Most minerals are chemical compounds of two or more elements; a few consist of only a single element. Almost all minerals are inor-

Figure 1. Pyrite from Huanzala, Huanuco, Peru. Pyrite is one the most common minerals and often forms well-developed crystals. Subject: 60 × 87 mm.

ganic, but a few consist of hydrocarbons or similar organic compounds.

Minerals normally occur assembled in groups or associations. They can be present in larger regional areas as rocks, or less extensively in veins, cavities, thin layers, incrustations, etc.

Geologists divide rocks into 'hard' and 'soft' rocks. The 'hard' rocks include the igneous and metamorphic rocks that are formed by processes in the deeper parts of the earth's crust, whereas the 'soft' rocks, the sediments, are largely produced by the erosion of rocks on the surface of the earth. Most 'hard' rocks consist of intergrowths of a few predominant minerals and a number of accessory minerals. Such a mineral association tells about a geological event that can be understood once the process of formation of the individual minerals is known.

A mineral will crystallize when the conditions during its formation—temperature, pressure, acidity, etc.— are within certain limits. If one of more of these conditions changes, the mineral may no longer be stable and in principle will be transformed into another mineral. Energy is used to break existing chemical bonds and reorganize or change the chemical content; and if enough energy—and time—are available the mineral will adjust to the new conditions and be changed into another mineral.

The amount of energy available is not, however, always sufficient to alter a mineral association significantly. In consequence, minerals formed under pressure and temperature conditions corresponding to a depth of 4 km below the surface of the earth can continue to exist in near-surface environments for many millions of years when parts of the earth's crust a kilometre or so thick are eroded and the conditions changed completely.

In other mineral associations changes in conditions do result in the alteration of one mineral to another. Typical examples are seen in copper deposits. The original ore minerals,

Figure 2. Quartz from Schmirnerthal, Tirol, Austria. Quartz is often found as crystals. A crystal is a body bounded by plane faces that is an external expression of an internal order. A mineral is a crystalline solid when it possesses this internal order, whether or not it is limited by plane faces. Field of view: 16 x 24 mm.

Figure 3. Sulphur (yellow) on calcite (white) from Conil, Andalusia, Spain. Crystals with well-developed faces are formed when the surroundings provide enough space. Field of view: 29 × 40 mm.

such as chalcopyrite, become unstable in the upper parts of the deposit and under the influence of circulating solutions with high oxygen contents they are transformed into minerals like malachite or azurite.

Some four thousand minerals are known on the earth. Each of them tells a story. Each is a unique combination of a given chemical compound and a given crystal structure formed under certain conditions. Within this multiplicity, many minerals are related, either because they consist of the same chemical compound, or because they have the same type of crystal structure. Quartz and cristobalite, for example, are two minerals with the same formula, SiO_2, but with different crystal structures; calcite ($CaCO_3$) and siderite ($FeCO_3$) have the same type of crystal structure, but differ in composition. Other minerals are unique, either because they con-

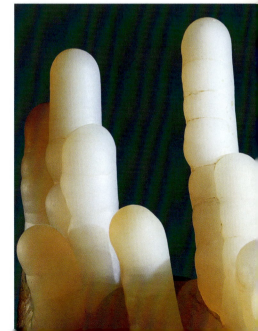

Figure 4. Chalcedony from the Faeroe Islands. Chalcedony is not crystalline in the normal sense but is built up of extremely small fibres. It typically occurs in stalactitic forms. Field of view: 102 × 131 mm.

Figure 5. Gneiss from Zackenberg, Greenland. Gneiss is a rock consisting of feldspars, quartz, and dark minerals such as biotite; the minerals are arranged in layers, giving the rock a foliated appearance. Gneiss is formed by alteration of other rocks at high pressure and temperature. Subject: 107 x 230 mm.

sist of an unusual combination of elements or because they have a rare type of crystal structure, or perhaps both. Such minerals are usually formed under exceptional circumstances.

Sorensenite, for instance, is a very special tin-containing silicate that has been found only in the Ilímaussaq complex in Greenland, although there it is relatively abundant.

Figure 6. Rhomb porphyry from Bulbjerg, Jutland, Denmark. A rhomb porphyry is an igneous rock with rhombic grains of feldspar in a fine-grained matrix. The feldspar grains were formed in the magma chamber before the magma, the melted mass of rock, was extruded to a higher level in the earth's crust or even to the surface, where the remaining part rapidly crystallized to a fine-grained mass. The rock illustrated is from a region near Oslo, Norway. Samples of it were brought to Denmark by glaciers during the last ice age. Subject: 73×79 mm.

Figure 7. Coarse-grained nepheline-syenite from Kangerdluarsuk, the Ilímaussaq complex, Greenland. The rock consists mainly of feldspar (white), nepheline (greyish white), arfvedsonite (black), astrophyllite (yellow brown), and several rare minerals that are not visible. These minerals were formed from a magma with an exceptional composition that was deficient in silica and had a surplus of sodium, titanium, zirconium, and several other rare elements. Its chemical composition provided appropriate circumstances for the development of many rare minerals. Subject: 170 × 184 mm.

Mineral names

Names of minerals and rocks normally end on *ite* or *lite*, from the Greek word *lithos*, stone. A few mineral names are very old, predating the Christian era, and their origins are generally obscure. Other names originate from old mining terms, especially German. For example, the endings *-spat* and *-blende* respectively signify minerals with good cleavage and lacking valuable metals. The vast majority of mineral names were established within the past two centuries. During the past few decades there has been a considerable increase in the rate, and at the present time about fifty new minerals are being recorded each year. The main reason for this recent ex-

Figure 8. Azurite from Bisbee, Arizona, USA. Azurite is a Cu-containing mineral found in the upper parts of Cu deposits, where it is formed by the alteration of Cu ore minerals such as chalcopyrite.
Field of view: 19 × 25 mm.

plosion is that better methods of investigation have made it possible to demonstrate a greater variety of mineral species.

In the beginning, minerals were often named according to a typical property. Axinite, for example, was named after the Greek word for axe, *axine*, because its crystals are typically shaped like the head of an axe. Astrophyllite was named by combining the Greek words *astron*, brilliance, and *pylon*, leaf, because of its high lustre and markedly foliated nature. Minerals were also named according to their chemical composition; for instance, chromite, which contains the element chromium. Later, as these possibilities gradually became exhausted, minerals were named after geographical locations, whether after a type locality (i.e. the place from which the mineral was first described) or after a landscape, county, or region in which the type locality was situated. Aragonite (Aragon, Spain), spessartine (Spessart, Germany), and

Figure 9. Axinite from Puiva, Urals, Russia. The name axinite is derived from the Greek word 'axine' and refers to the characteristic shape of crystals of the mineral, which resemble the head of an axe. Subject: 36 × 57 mm.

Figure 10. Spessartine from Nathrop, Colorado, USA. Spessartine is named after the type locality near Spessart, Bayern, Germany. (The type locality of a mineral is the place from which it was first described.) Field of view: 11 × 17 mm.

Figure 11. Phillipsite from Gads Hill, Tasmania, Australia. Phillipsite is named after the English mineralogist William Phillips (1775–1828). Field of view: 64 × 90 mm.

elbaite (Elba, Italy) are examples. It later also became common to name minerals after individuals, especially chemists and mineralogists: phillipsite was named after the English mineralogist W. Phillips and ussingite after the Danish geologist N.V. Ussing.

Contemporary practice in naming minerals generally follows the same principles as in the past, but today the new mineral and its proposed name have to be approved by an international commission before publication.

Figure 12. Apatite with phlogopite from Ze Pinto Prospect, Aldeia, Minas Gerais, Brazil. Subject: 66 × 78 mm.

Crystallography

Almost all minerals are crystalline. A crystalline material is a solid in which the chemical constituents are ordered in a particular three-dimensional pattern. Under favourable conditions the material—or crystal—is bounded by plane faces, but most crystals are formed under conditions that do not favour the development of crystal faces. Even so, the material is nevertheless crystalline because it possesses the fundamental property of having an ordered internal structure.

The crystalline state is fundamentally different from the gaseous and the liquid states, in which the chemical constituents are distributed in a disordered fashion. A few materials, such as glass (which is generally regarded as a solid) are not crystalline. Glass can be regarded as a supercooled liquid that did not have time to crystallize during the cooling process. It is an unstable phase that will slowly transform into a crystalline solid. Glass and other non-crystalline materials are *amorphous*; i.e. they are without any form.

There is a fundamental difference between crystalline and amorphous materials. The chemical and physical properties of amorphous materials are uniform in all directions, but the properties of crystalline materials can vary according to the direction in which they are measured. This is evident, for example, in optical properties, cleavage, and hardness, as will be shown later.

Crystallography is the branch of science that deals with crystals, their forms and symmetry, their internal structure, and the chemical and physical properties that are determined by their structure. It is thus a fundamental part of mineralogy. Some basic elements of crystallography are outlined below. Real crystals are usually incomplete, but for explanatory purposes we shall consider perfect crystals as models.

Crystal geometry

Crystal geometry or *crystal morphology* deals with crystals and their symmetry, faces, and forms, as well as with the terminology that is used to describe crystals.

When we colloquially say that something is symmetrical we usually mean that it has a certain degree of regularity or harmony, but we do not necessarily consider a specific type of symmetry. Most often, though, we think of something bilateral, such as the human body. In crystallography we must expand and make concrete the concept of symmetry. We can begin with the following classification of symmetry elements:

(1)   symmetry in relation to a plane: a *plane of symmetry*;
(2)   symmetry in relation to a point: a *centre of symmetry*;
(3)   symmetry in relation to an axis: an *axis of symmetry*;

(4)   symmetry in relation to rotary inversion: an *inversion axis*.

The action described by a symmetry element—a reflection, rotation, etc.—is called a *symmetry operation*.

Plane of symmetry

A plane of symmetry divides a crystal into two halves, one half being a mirror image of the other. A crystal can be divided into halves by many planes, but only planes dividing it into two mirror images are planes of symmetry. Crystals can have one or several planes of symmetry, or none. A plane of symmetry, also called *a mirror plane*, is designated by an *m* in the symbols for the 32 crystal classes described below.

Centre of symmetry

A centre of symmetry is a point in the crystal in which a face can be 'mirrored' to a corresponding (inverse) face that is diametrically

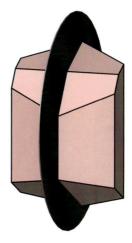

Figure 13. A plane of symmetry divides a crystal into parts that are mirror images of each other.

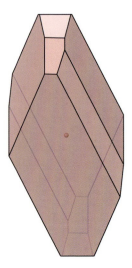

Figure 14. A centre of symmetry, or point of inversion, generates for every face a corresponding one diametrically opposite and in a reversed position.

Axes of symmetry

An axis of symmetry is a line or axis through a crystal about which the crystal can be rotated to bring it into an identical position a number of times in the course of one revolution. If during the course of a full rotation of 360° about an axis of symmetry the crystal is brought into an identical position six times—or once in every 60°—the axis is a *sixfold axis*. If the angle between successive identical positions is 90°, the axis is a *fourfold axis*; if the angle is 120°, it is a *threefold axis*; and if the angle is 180° it is a *twofold axis*. If the crystal is in an identical position only once during a complete 360° rotation, we can speak of a *onefold axis*. Every crystal has in fact an infinite number of onefold axes, but the concept does make sense, as will be shown below. The symbol for an axis of symmetry is the number of identical positions reached in the course of a full rotation, e.g. 6. A crystal can have several symmetry axes. If it has a sixfold axis and a number of twofold axes, the sixfold axis is called the *principal*

opposed to it. On a well-developed crystal with a centre of symmetry every face has a corresponding face that is diametrically opposite and parallel to it. Some crystal forms, such as the tetrahedron, have no centre of symmetry. A centre of symmetry is also called a *point of inversion*.

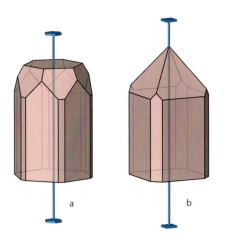

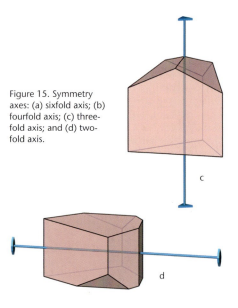

Figure 15. Symmetry axes: (a) sixfold axis; (b) fourfold axis; (c) threefold axis; and (d) twofold axis.

Figure 16. A $\bar{3}$ inversion axis repeats a face by a symmetry operation consisting of a rotation of 120° followed by a reflection over a point of inversion.

try or point of inversion. The following inversion axes are possible: $\bar{1}$ (read as 'bar one') with a rotation of 360°, equivalent to a centre of symmetry; $\bar{2}$ with a rotation of 180°, identical to a symmetry plane; $\bar{3}$ with a rotation of 120°, equivalent to a threefold axis combined with a centre of symmetry; and $\bar{6}$ with a rotation of 60°, identical to a threefold axis perpendicular to a mirror plane. A $\bar{4}$ inversion axis with a rotation of 90° is a unique symmetry element that is the only possible way of describing a given combination of faces.

Combination of symmetry elements

Symmetry elements can be combined in various ways: several planes of symmetry; several axes of symmetry; axes of symmetry with planes of symmetry, and so forth. The possibilities are limited, however, because the symmetry elements influence each other. Figure 18 shows an example from a tetragonal crystal. Suppose that the axis labelled a is a twofold axis. The axis labelled b then also has to be a twofold axis because there is a fourfold axis perpendicular to the plane of the paper, i.e. to the a and b axes. In addition,

axis and is placed at the beginning of the symbol for the crystal class (see below). Other axes of symmetry, such as a fivefold axis, are not possible.

Inversion axes

An inversion axis is a symmetry element that can be explained as a symmetry operation that combines a fractional rotation (i.e. one of less than 360°) with a subsequent reflection through an imaginary centre of symme-

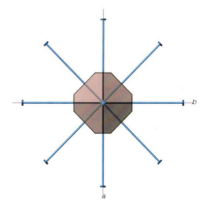

Figure 17. A $\bar{4}$ inversion axis repeats a face by a symmetry operation consisting of a rotation of 90° followed by a reflection over an imaginary point of inversion.

Figure 18. Symmetry elements influence one other: if, e.g., there is a twofold axis perpendicular to a fourfold axis, three other twofold axes perpendicular to the fourfold axis are automatically created.

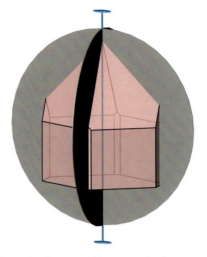

Figure 19. When a crystal has two, and only two, symmetry planes they have to be perpendicular to each other, and a twofold axis is automatically formed at the line of intersection.

symmetry elements will otherwise inevitably arise. The mutual dependence is further illustrated by a crystal with only two planes of symmetry: they must be perpendicular to each other and will thus automatically create a twofold axis of symmetry at their line of intersection (Figure 19).

In all, 32 combinations of symmetry elements are possible. They are called *crystal classes* or *point groups* and are grouped into seven *crystal systems* (Figure 20).

The unit cell

There exists a multiplicity of crystal faces and combinations of faces, and we need to establish principles for describing the external appearance of a crystal. Such principles are best illustrated by taking a unit cell as a starting point.

A unit cell is like a box that by virtue of its chemical content and its form, i.e. its dimensions and symmetry, defines a mineral. The unit cell is thus in effect the smallest unit in a crystal, which can be conceived as a gigantic stacking of unit cells in three-dimensional space without leaving any gaps. The stacking can take place in various ways and can thus result in different crystal forms, all having the same symmetry as the unit cell, and with angles between faces that can be de-

this set of twofold axes will automatically create another set of twofold axes 45° from the first and in the same plane. Thus, there are in all four twofold axes perpendicular to the fourfold axis.

The mutual dependence of the symmetry elements is also demonstrated in a combination of a single plane of symmetry with a single axis of symmetry. This is possible only if the axis is perpendicular to the plane; new

Crystal system	The 32 crystal classes
Cubic	$4/m\overline{3}2/m$, $2/m\overline{3}$, $\overline{4}3m$, 432, 23
Hexagonal	$6/m2/m2/m$, $\underline{6}mm$, $6/m$, 6, 622, $\overline{6}m2$, $\overline{6}$
Trigonal	$\overline{3}2/m$, $3m$, 32, $\overline{3}$, 3
Tetragonal	$4/m2/m2/m$, $4/m_{\perp}$, 4, $\overline{4}2m$, 4mm, 422, $\overline{4}$
Orthorhombic	$2/m2/m2/m$, 2mm, 222
Monoclinic	$2/m$, m, 2
Triclinic	$\overline{1}$, 1

Figure 20. There are in all a total of 32 possible combinations of symmetry elements. They are called *crystal classes* or *point groups*. The meanings of the symbols are as follows: $4/m$ (read as 'four over m'), a fourfold axis and perpendicular to this a plane of symmetry; $6mm$ (read as 'six m m'), a sixfold axis and two sets of planes of symmetry, i.e. a total of six. When the oblique stroke (/) is not included, it signifies that the axis is not perpendicular to the plane but is contained in the plane. 32 (read as 'three two'), denotes a threefold axis and, perpendicular to this, a set of twofold axes, i.e. three twofold axes. A bar over a symmetry axis designates the axis as an inversion axis; e.g. $\overline{3}$ (read as 'bar three'). The sequence of the symbols is significant; e.g. $3m$ is a trigonal class with a threefold axis as the principal axis, whereas $m3$ is a cubic class, m before the 3 referring to the 'axial planes' described in the text below.

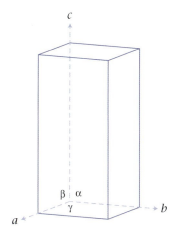

Figure 21. The dimensions of a unit cell are characterized by the three edge lengths and the three angles between the edge directions. The unit cell shown here is orthorhombic, with three edges of different lengths at right angles.

duced from the dimensions of the unit cell.

Figure 21 shows the unit cell of sulphur. It measures $10 \times 13 \times 25$ Å (ångstrom units; 1 Å $= 10^{-10}$ m) and the angles between adjacent edges are all $90°$. Sulphur has orthorhombic symmetry, which includes a centre of symmetry, three planes of symmetry perpendicular to each other, and three twofold axes of symmetry at the intersections of the planes. The three edges of the unit cell indicate a system of coordinates, from which the six dimensions that define the unit cell can be determined: the edge-lengths a, b, and c and the angles between the edges α, β, and γ. In sulphur these dimensions are: 10, 13, and 25 Å respectively and $\alpha = \beta = \gamma = 90°$.

The three directions represented by a, b, and c also indicate the *crystallographic axes*, which coincide with the three twofold axes of the crystal.

A system of coordinates with three axes is used for all seven crystal systems. The intercepts and interfacial angles vary from system to system, and the following relationships apply:

Triclinic system: $a \neq b \neq c$, $\alpha \neq \beta \neq \gamma \neq 90°$.
Monoclinic system: $a \neq b \neq c$, $\alpha = \gamma = 90°$, $\beta \neq 90°$.

Orthorhombic system: $a \neq b \neq c$, $\alpha = \beta = \gamma = 90°$.
Tetragonal system: $a = b \neq c$, $\alpha = \beta = \gamma = 90°$.
Cubic system: $a = b = c$, $\alpha = \beta = \gamma = 90°$.
Trigonal system: $a = b = c$, $\alpha = \beta = \gamma \neq 90°$.
Hexagonal system: $a = b \neq c$, $\alpha = \beta = 90°$, $\gamma = 120°$.

For hexagonal and trigonal crystals a system of coordinates with four axes is also used since it better represents the symmetrical relationships.

Crystal faces

Figure 22 shows a number of unit cells stacked to form a three-dimensional lattice of points. Every lattice point is a place at which the corners of eight unit cells meet. The point lattice represents the internal structure of the crystal: the unit cell dimensions, the distances, and the angles are integral elements of the point lattice. The planes delimited by lattice points are called *lattice planes*, and they represent possible crystal faces. Only lattice planes can be crystal faces, which explains the important fact that for a particular mineral there are specific orientations of crystal faces and, correspondingly, specific angles between crystal faces. Figure

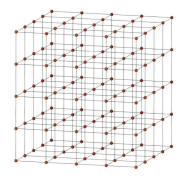

Figure 22. Stacking of unit cells where their corners form a point lattice. Planes containing lattice points are possible crystal faces.

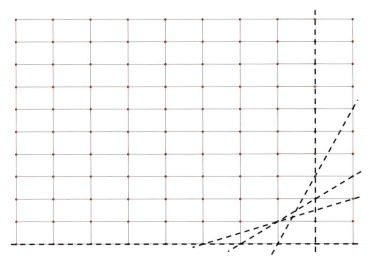

Figure 23. In a two-dimensional net of lattice points only certain directions are possible for a line (or, in three dimensions, a plane) that has to intersect lattice points.

23 illustrates in two dimensions the principle that for a given net of lattice points only certain directions of lines are possible if these lines are to intersect lattice points. Planes are similarly restricted if they are to intersect lattice points in three dimensions.

The planes intersecting the closest net of lattice points usually form the most prominent crystal faces.

Designation of crystal faces

Crystal faces are identified by specific codes. These codes, called the *Miller indices*, consist of three numerals (indices) in round brackets, e.g. (100), (110), or (321). The first numeral refers to the *a* axis, the second to the *b* axis, and the third to the *c* axis. When the values are unknown the indices are simply written as (*hkl*).

This method of designating crystal faces is based on the geometrical fact that the intercepts made by various crystal faces on the crystallographic axes are related to each other by simple ratios. This is a consequence of the regularity of a crystalline structure, in which only certain directions are potential crystal faces.

A plane cutting all three axes is chosen as a plane of reference for specifying the indices of crystal faces. This plane is designated as the *unit plane*, and the intercepts it makes on the three axes become units for indexing other faces. The principle is depicted in Figure 24, in which possible faces are drawn on a point lattice. The plane $A_1B_1C_1$ is chosen as the plane of reference, and thus its intercepts on the *a*, *b*, and *c* axes are reference units *a*, *b*, and *c* respectively.

Every other crystal face can now be described by the intercepts it makes on the three axes when related to the intercepts of the unit plane. This is done in the following way:

intercept on the *a* axis = a/h,
intercept on the *b* axis = b/k,
intercept on the *c* axis = c/l,

where *a*, *b*, and *c* are the intercepts of the unit plane and *h*, *k*, and *l* are simple integers or zero. The unit plane itself has the symbol (111), since its intercepts are exactly *a*, *b*, and *c*. When inserted in the equations, these intercepts give the value of 1 to *h*, *k*, and *l*. Now the plane $A_1B_2C_2$ (Figure 24) has the indices

Figure 24. A point lattice with two possible crystal faces shown. The green plane ($A_1B_1C_1$) is chosen as a unit plane; it thus has the indices (111). The blue plane ($A_1B_2C_2$) then has the indices (211). For further explanation, see text.

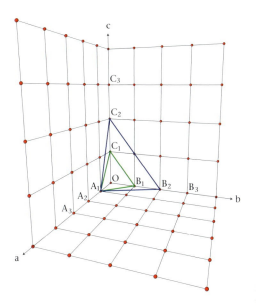

$A_1 - O = a\quad = a/h,$
$B_2 - O = 2b = b/k,$
$C_2 - O = 2c = c/l,$

where hkl become 1, ½, ½. However, as we are dealing with ratios the fractions can be cleared by multiplying by two, giving the symbol (211). This is read as *two one one* and is normally written without commas.

The indices for other faces are readily derived: e.g. $A_1B_3C_3$: (311) and $A_2B_1C_1$: (122). The plane containing A_1 and B_2 and parallel with the c axis has $a = a/h$, $2b = b/k$ and ∞ (infinity) $= c/l$, giving the indices 1,½,0, which on clearing fractions gives (210).

A face cutting only the a axis and thus parallel with the other two axes is called (100), and in the same way we obtain (010) and (001). An intercept at the negative end of an axis is marked with a bar over the relevant index: for example ($1\bar{1}0$).

Indices are generally written as precisely as possible; when numerical values are known they are included, e.g. (320). If the actual values were not known the indices in the previous example could be written simply as ($hk0$). In some crystal forms the exact values may not be known although their mutual relationship is known. A form might, for instance, be written as (hhl), indicating that $h = k$.

For hexagonal and trigonal crystals another notation is usually preferred, as described under the hexagonal crystal system.

Crystal forms

The word *form* has a special meaning in crystallography. A *crystal form* is an assemblage of faces that are identically related to the symmetry elements.

As explained above, a face is defined by a set of indices, (hkl), that describe the orientation of the face in relation to the crystallographic axes. For example, a pyramid face, (111), on an orthorhombic crystal will be repeated seven times by virtue of the symmetry elements of the orthorhombic system. These eight faces, identical with respect to the symmetry elements, constitute a crystal *form*; in this case the orthorhombic bipyramid. Whereas individual faces are enclosed in normal (round) brackets, the form is enclosed in braces ('curly brackets'): {111}.

Square brackets, [001], have a different meaning. They are used to indicate a direction in the crystal: in this instance the direction of the c axis.

Forms may be *open* or *closed*. A cube, {100}, encloses space and can occur on its own, whereas a monoclinic prism, {$hk0$}, is an open form and must be combined with another form to enclose space.

Many crystal forms are named after the number of faces or after the shape of the faces when they appear alone on a crystal. A rhombic dodecahedron is thus a crystal form with twelve rhombic faces (*dodeca*, Gk. 'twelve'; *hedron*, Gk. 'face').

A crystal form is a *special form* when it has a special orientation with respect to the symmetry elements of the crystal. Its faces can be parallel with a set of planes of symmetry or perpendicular to an axis of symmetry. A crystal form is a *general form* when its faces have a general, non-specific, or 'oblique' orientation with respect to the symmetry elements. A general form has the largest possible number of faces in a crystal class.

The cubic system

The cubic system, also known as the *isometric system*, includes five crystal classes, all characterized by having four threefold axes of symmetry, although the remaining symmetry elements vary (Figure 20). The hexoctahedral class, $4/m\bar{3}2m$, has the highest cubic symmetry and is accordingly the class with the highest possible symmetry of all. Generally speaking, the crystal class having the highest possible symmetry in a system is called the *holohedral class*; so $4/m\bar{3}2m$ is the cubic holohedral class. We shall concentrate on this class and shall mention only briefly two others: $\bar{4}3m$ and $m3$.

The hexoctahedral class, $4/m\bar{3}2/m$

Figure 25 illustrates the many symmetry elements of the hexoctahedral class. There are three fourfold axes of symmetry, four threefold inversion axes, and six twofold axes. As seen on the cube, the fourfold axes pass through the centres of two opposite faces and are perpendicular to each other; the threefold axes pass through one corner to the diametrically opposite corner; and the twofold axes pass through the midpoints of opposite edges. There are nine planes of symmetry in all, three parallel to the cube faces and

perpendicular to each other (the 'axial planes') and six planes each containing a pair of opposite edges ('diagonal planes'). Finally, the class has a centre of symmetry. As mentioned above, a threefold axis combined with a centre of symmetry is equal to a threefold inversion axis. Accordingly, the complete symmetry symbol of this class is $4/m\bar{3}2/m$.

In order to represent the individual crystal forms a set of crystallographic axes has to be selected. For the cubic system the three fourfold axes of symmetry are an obvious choice. The three axes are of equal length; thus, $a = b = c$.

Crystal forms:

Cube, {100}. A face perpendicular to a fourfold axis and thus parallel to the other two fourfold axes has a special relationship to the symmetry elements. The symbols for the face are (100), or $(\bar{1}00)$ if it cuts the axis at the negative end. Despite the many symmetry elements of this class, the face (100) is repeated only five times: to (010), $(0\bar{1}0)$, (001), $(00\bar{1})$, and $(\bar{1}00)$. These six faces together constitute the well-known crystal form the *cube*, or *hexahedron* (= six faces), with the

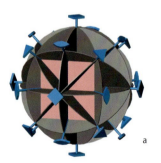

a

b

Figure 25. The crystal class $4/m\bar{3}2/m$: A cube, {100}, (a) with and (b) without information about symmetry elements. The centre of symmetry is not shown.

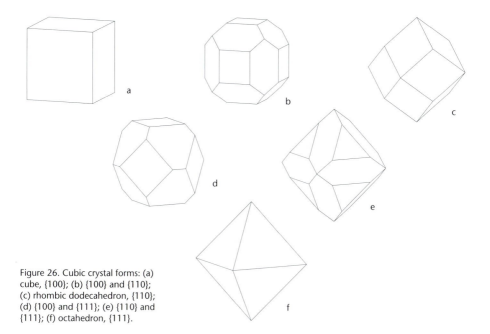

Figure 26. Cubic crystal forms: (a) cube, {100}; (b) {100} and {110}; (c) rhombic dodecahedron, {110}; (d) {100} and {111}; (e) {110} and {111}; (f) octahedron, {111}.

form {100}. As for all cubic crystal forms, the cube is a closed form.

Octahedron, {111}. A face perpendicular to a threefold axis will make intercepts of equal lengths on the three axes and will have the symbol (111) when cutting the axes at their positive ends. The face is repeated seven times by the symmetry elements and together with these constitutes an octahedron (= eight faces), {111}.

Rhombic dodecahedron, {110}. A face perpendicular to a twofold axis makes equal intercepts with two of the crystallographic axes and is parallel with the third, resulting in the symbol (110). It is repeated eleven times to give a rhombic dodecahedron (*dodecahedron* = twelve faces), {110}.

The three forms cube, octahedron, and rhombic dodecahedron will have square, equilateral triangular, and rhombic faces respectively if they appear alone on a crystal. Figure 26 shows combinations in which the sizes and shapes of the crystal faces vary. These combinations clearly illustrate that a crystal face is not to be recognized by its outline but by its orientation with respect to the symmetry elements. A face of a cube is recognized by being normal to a fourfold axis of symmetry; a face of an octahedron by being normal to a threefold axis of symmetry; and a face of a rhombic dodecahedron by being normal to a twofold axis of symmetry.

The cube, octahedron, and rhombic dodecahedron are special crystal forms because they have a particular orientation with respect to the symmetry elements. There are other special crystal forms:

Tetrahexahedron, {hk0}, whose faces are parallel to one fourfold axis of symmetry and make unequal intercepts with the other two fourfold axes. The crystal form has 24 faces and its symbols are {210}, {310}, or otherwise according to the ratios of the intercepts on the two axes.

27

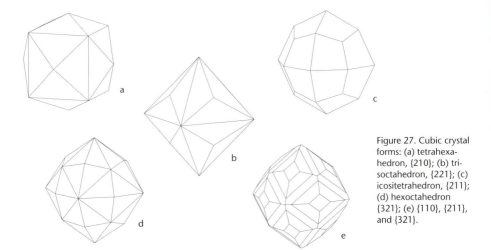

Figure 27. Cubic crystal forms: (a) tetrahexa-hedron, {210}; (b) tri-soctahedron, {221}; (c) icositetrahedron, {211}; (d) hexoctahedron {321}; (e) {110}, {211}, and {321}.

Trisoctahedron, {hhl}, also has 24 faces. The special notation {hhl} implies that h = k and h > l. The form could be {221} or {331}.

Icositetrahedron, {hkk}, also has 24 faces. In this case the notation implies that h > k = l.

The icositetrahedron could be {211}. An ico-sitetrahedron is also called *a trapezohedron*.

Hexoctahedron, {hkl}. In contrast to the special forms, this is a general crystal form with faces that do not have a particular orientation in

Figure 28. Grossular with {110} from Ocna de Fier, Romania. Field of view: 27 × 45 mm.

relation to the symmetry elements. As a result this face is repeated to give a maximum of 48 faces. It cuts the three axes in intercepts of different lengths, and the symbol is therefore {hkl}, e.g. {321}. The general form gives its name to the class.

Metals such as gold, silver, and copper belong to this class. Diamond, halite, fluorite, galena, magnetite, and garnets are other well-known examples.

The hextetrahedral class, $\bar{4}3m$

As compared with the holohedral class, the hextetrahedral class has lost some symmetry elements. Three planes of symmetry (the axial planes) are absent, as are the six twofold axes; and the fourfold axes are reduced to fourfold inversion axes, $\bar{4}$. The absence of these symmetry elements has no effect on {100} and {110}, which are still the cube and rhombic dodecahedron, but in this class {111} is a *tetrahedron* (= four faces). There are two possible orientations of a tetrahedron with respect to the axes, a positive, {111}, and a negative, {1$\bar{1}$1}. The *tristetrahedron*, {hkk},

Figure 29. Magnetite with {111}, {311}, and small faces belonging to {110}, from the Gardiner complex, Greenland. Field of view: 44 × 75 mm.

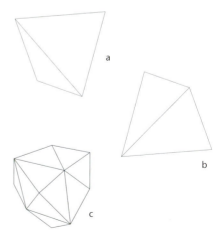

Figure 30. The crystal class $\bar{4}3m$: (a) positive tetrahedron, {111}; (b) negative tetrahedron, {1$\bar{1}$1}; (c) hextetrahedron, {321}.

Figure 31. Tetrahedrite, dominated by {211}, from Horhausen, Rheinland-Pfalz, Germany. Field of view: 58 × 87 mm.

e.g. {211}, is another special form. The general form is a *hextetrahedron* (= 6 × 4) {*hkl*} or {*hkl̄*} consisting of 24 faces. Sphalerite and tetrahedrite, the latter being named after the crystal form, are well-known mineralogical examples.

The didodecahedral class, 2/m3̄

In the didodecahedral class the six 'diagonal planes of symmetry' and the six twofold axes are absent, and the three fourfold axes are reduced to twofold axes. The 'axial planes of symmetry' and the threefold inversion axes are still present. The cube, {100}, octahedron, {111}, and rhombic dodecahedron, {110}, are found in this class as well, whereas the form {*hk*0}, which in the holohedral class is a tetrahexahedron, is in this class a *pentagonal dodecahedron* (= twelve pentagonal faces), e.g.

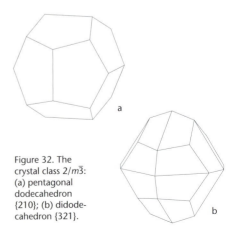

Figure 32. The crystal class 2/m3̄: (a) pentagonal dodecahedron {210}; (b) didodecahedron {321}.

{210}; the form is also called *a pyritohedron* after pyrite. The general form {*hkl*} is *a didode-cahedron* (2 × 12 faces), also known as *a diploid.*

Pyrite is the best-known example of this class. The class offers good examples of crystals apparently displaying higher symmetry than they in fact possess. Thus, the cube appears to be holohedral with all possible cubic symmetry elements, but it is actually found in all five cubic classes. When the cube is the only form present it is thus not possible to allocate a crystal to a particular cubic class. In order to do that, the crystal must show the general form, or alternatively a combination of special forms that unequivocally point to a specific class.

Pyrite is often found as cubes. Striation of cube faces can reveal that the symmetry is not holohedral (Figure 33). The striae demonstrate the lack of fourfold axes. They are caused by oscillatory growth of the two crystal forms {100} and {*hk*0}.

The tetragonal system

The tetragonal crystal system includes seven classes. All classes have a fourfold axis of symmetry that is referred to as the principal axis. This axis can exist alone or in combination with a centre of symmetry, planes of symmetry, and/or twofold axes. Here we consider the holohedral class, $4/m2/m2/m$, and briefly mention the two classes $4/m$ and $\bar{4}2m$.

The ditetragonal bipyramidal class, $4/m2/m2/m$

The symmetry content of the ditragonal bipyramidal class is: (i) a fourfold axis, the principal axis, and perpendicular to it a plane of symmetry; (ii) four twofold axes lying in

Figure 33. Pyrite with {210} from Huanzala, Huanuco, Peru. The faces are striated owing to oscillilatory growth between the two forms {210} and {100}. Subject: 31 × 48 mm.

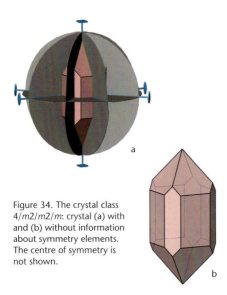

Figure 34. The crystal class $4/m2/m2/m$: crystal (a) with and (b) without information about symmetry elements. The centre of symmetry is not shown.

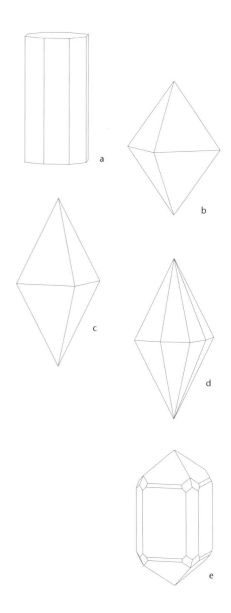

this plane of symmetry, two by two perpendicular to each other and with 45° between the pairs; (iii) four planes of symmetry, each including the principal axis and a twofold axis; and (iv) a centre of symmetry (Figure 34). The complete symbol is $4/m2/m2/m$. A rectangular system of coordinates is chosen as the set of axes. The principal axis is the c axis, and a pair of twofold axes are the a and b axes. It follows from the presence of a fourfold axis that the a and b axes are of equal

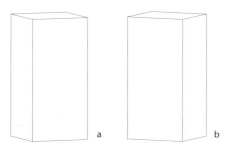

Figure 35. Tetragonal crystal forms: (a) first-order tetragonal prism, {110}, and pinacoid, {001}, and (b) second-order tetragonal prism, {100}, and pinacoid, {001}.

Figure 36. Tetragonal crystal forms: (a) ditetragonal prism, {210}, and a pinacoid, {001}; (b) first-order tetragonal bipyramid, {hhl}; (c) second-order bipyramid, {h0l}; (d) ditetragonal bipyramid, {hkl}; (e) zircon with {100}, {110}, {101}, {301}, and {211}.

length: $a = b \neq c$. As there are two sets of two-fold axes, there are two orientations that are equally valid. Once an orientation is selected it should be kept. The principal axis is always taken as the vertical axis.

Crystal forms:

Pinacoid, {001}. A face perpendicular to the fourfold axis is repeated once only to produce a crystal form consisting of the two faces (001) and (00$\bar{1}$).

Tetragonal prism, {110} or {100}. A face parallel to the fourfold axis and perpendicular to a twofold axis is repeated to make a four-faced prism with a square cross-section. Since there are two possible orientations of the crystal, the prism could be {110} with equal intercepts on the two horizontal axes, or {100} with the prism faces perpendicular to the horizontal axes. They are called a first-order prism, {110}, and a second-order prism, {100}. The two prisms can both be present in the same crystal.

Ditetragonal prism, {hk0}. A face parallel with the fourfold axis of symmetry and cutting the horizontal axes in intercepts of different

Figure 37. Cassiterite from Xue Bao Diang, Sitchuan, China. Subject: 53 × 76 mm.

Figure 38. Wulfenite from St. Anthony Mine, Mammoth–Tiger District, Pinal County, Arizona, USA. Field of view: 10 × 18 mm.

lengths is repeated seven times to form a ditetragonal prism. The form has eight faces: two (*di*) by four (*tetra*).

Tetragonal bipyramid, {*hhl*} or {*h0l*}. A face cutting both the principal axis and the horizontal axes and perpendicular to one of the vertical symmetry planes is repeated to form one pyramid on the upper half of the crystal and another pyramid on the lower half of the crystal, which together form a tetragonal bipyramid. The form consists of eight faces and, like a prism, it can be notated in two orientations, {*hhl*}, or {*h0l*}.

Ditetragonal bipyramid, {*hkl*}. A face with a general orientation is repeated to form a ditetragonal bipyramid with sixteen faces. (Two (*di*) by four (*tetra*) by two (*bi*) equals sixteen.) Cassiterite, rutile, zircon, and apophyllite are examples of minerals in this class.

The tetragonal bipyramidal class, 4/*m*

In the tetragonal bipyramidal class the twofold axes are absent, as are the vertical symmetry planes. A fourfold axis perpendicular to a symmetry plane remains, plus a centre of symmetry. The special holohedral crystal

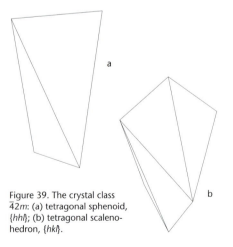

forms also occur in this class, with the exception of {*hk0*}, which is reduced to a *tetragonal prism*. This is due to the absence of the twofold axes and the vertical symmetry planes, which also lead to the general form, the *tetragonal bipyramid*, {*hkl*}, having only eight faces. Wulfenite, scapolite, and scheelite belong to this class.

The tetragonal scalenohedral class, $\overline{4}2m$

In the tetragonal scalenohedral class the principal axis is reduced to a fourfold inversion axis, the horizontal plane of symmetry is absent, and there is only one set of twofold axes and one set of vertical planes of symmetry. The open crystal forms of the holohedral class are also found here as well as a second-order *tetragonal bipyramid*, whereas the corresponding first-order form is four-faced, a *tetragonal sphenoid*, {*hhl*}. This form clearly demonstrates the fourfold inversion axis. The general form of this class is a *tetragonal scalenohedron* {*hkl*}. Chalcopyrite belongs to this class.

The hexagonal system

The hexagonal crystal system consists of seven crystal classes, all having a sixfold axis of symmetry as the principal axis. The hexagonal and trigonal crystal classes are sometimes united into one crystal system because of the close relationships between crystals with sixfold and threefold symmetries.

The dihexagonal bipyramidal class, 6/*m*2/*m*2/*m*

The dihexagonal bipyramidal class of the hexagonal system is the holohedral class

Figure 39. The crystal class $\overline{4}2m$: (a) tetragonal sphenoid, {*hhl*}; (b) tetragonal scalenohedron, {*hkl*}.

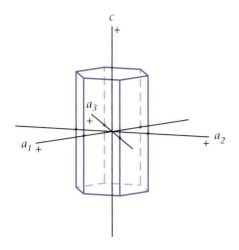

Figure 40. A set of four axes is used in the description of hexagonal and trigonal crystals.

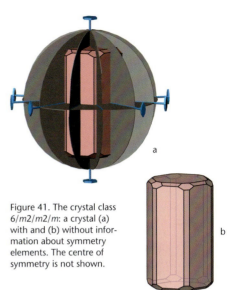

Figure 41. The crystal class $6/m2/m2/m$: a crystal (a) with and (b) without information about symmetry elements. The centre of symmetry is not shown.

with the following symmetry elements: (i) a sixfold axis of symmetry, and perpendicular to this a symmetry plane, 'the horizontal plane'; (ii) six vertical planes of symmetry that intersect at the sixfold axis; (iii) a twofold axis perpendicular to each vertical symmetry plane, and hence six twofold axes, which are regarded as two sets of three axes, each set being at 30° to the other; and (iv) a centre of symmetry. These symmetry elements are expressed in the symbol $6/m2/m2/m$.

The hexagonal and trigonal crystal forms are best described by using a set of four axes, as shown in Figure 40. There is for each system a principal axis and one of the two sets of twofold axes, giving three horizontal axes equal in length with their positive ends as indicated in the figure. The axes are labelled $a_1 = a_2 = a_3 \neq c$. As the other set of twofold axes can equally well be used, there are two orientations of equal status. Having four reference axes rather than three requires an additional index for a face. The fourth index is designated i and the complete indexes are ($hkil$), in which h refers to the a_1 axis, k to the a_2 axis, i to the a_3 axis, and l to the c axis. Note that the negative end of the a_3 axis points towards the observer. The geometrical relations imply that $h + k + i = 0$.

Crystal forms:

Pinacoid, {0001}. A face perpendicular to the principal axis is repeated once to give an open crystal form of two faces: (0001) and (000$\bar{1}$).

Hexagonal prism, {10$\bar{1}$0} or {11$\bar{2}$0}. Faces parallel to the principal axis and perpendicular to one set of the vertical symmetry planes are

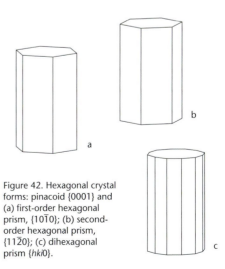

Figure 42. Hexagonal crystal forms: pinacoid {0001} and (a) first-order hexagonal prism, {10$\bar{1}$0}; (b) second-order hexagonal prism, {11$\bar{2}$0}; (c) dihexagonal prism {hki0}.

35

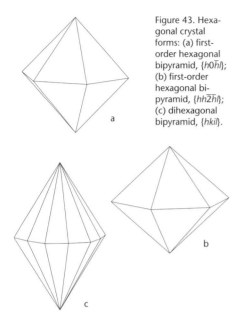

Figure 43. Hexagonal crystal forms: (a) first-order hexagonal bipyramid, {h0h̄l}; (b) first-order hexagonal bipyramid, {hh2̄hl}; (c) dihexagonal bipyramid, {hkil}.

assembled in a prism of six faces. Two possibilities exist in relation to the chosen set of axes: a first-order prism, {101̄0}, and a second-order prism, {112̄0}. For example, the prism {101̄0} consists of the following faces: (101̄0), (11̄00), (01̄10), (1̄010), (1̄100), and (011̄0).

Dihexagonal prism, {hki0}. A face parallel to the principal axis and not perpendicular to a vertical plane of symmetry is repeated eleven times to give a prism of 12 faces, a dihexagonal prism, {hki0}. The symbol is of the type {213̄0}.

Hexagonal bipyramid, {h0h̄l} or {hh2̄hl}. A face cutting the principal axis and perpendicular to a vertical plane of symmetry is repeated to give a hexagonal bipyramid with twelve faces: a single pyramid on the upper half and a corresponding one on the lower half. Two orientations are again possible. Known relations are included in the indexes. Thus, if a second-order hexagonal bipyramid has $h = k$, then $i = \overline{2h}$.

Figure 44. Beryl var. aquamarine from Nagar, Hunza Valley, Pakistan. Field of view: 104×156 mm.

Figure 45. Pyromorphite from Zvezdel-Ptcheloyad, Kurdzhali, Bulgaria. Subject: 71 × 90 mm.

Dihexagonal bipyramid, {hkil}. A face with a general orientation with respect to the symmetry elements is repeated to give a dihexagonal bipyramid consisting of 24 faces: two (*di*) by six (*hexa*) by two (*bi*) = 24. Beryl belongs to this crystal class.

The hexagonal bipyramidal class, 6/*m*

As compared with the holohedral class, the hexagonal bipyramid class lacks the twofold axes and the vertical planes of symmetry. Remaining are the principal axis, the plane of symmetry perpendicular to it, and a centre of symmetry. The general crystal form {hkil} is *a hexagonal bipyramid* of twelve faces. Apatite and pyromorphite belong to this class.

The hexagonal pyramidal class, 6

The hexagonal pyramidal class has only one symmetry element, a sixfold axis of symmetry; even the centre of symmetry is absent.

The axis is polar, i.e. the positive and negative ends of the axis are symmetrically distinct. The general crystal form {hkil} is *a hexagonal pyramid* of only six faces. Nepheline belongs to this class.

The trigonal system

The trigonal system includes five crystal classes with a threefold axis as the principal axis. As mentioned above, trigonal and hexagonal crystal forms are usually described by using a set of four reference axes (see Figure 40) to express the symmetry relationships.

The ditrigonal scalenohedral class, $\bar{3}2/m$

One might expect that the holohedral class—the class with the highest possible symmetry within a system—would also include a principal axis and perpendicular to it a plane of

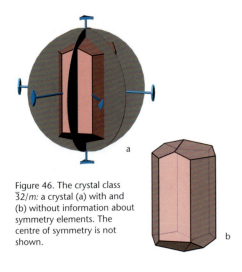

Figure 46. The crystal class $\bar{3}2/m$: a crystal (a) with and (b) without information about symmetry elements. The centre of symmetry is not shown.

fold axes perpendicular to the planes of symmetry; and a centre of symmetry. The threefold axis and centre of symmetry are combined to give a threefold inversion axis, resulting in the symbol $\bar{3}2/m$.

Crystal forms:

Pinacoid, {0001}, hexagonal prisms in a first orientation, {10$\bar{1}$0}, and a second orientation, {11$\bar{2}$0}; the dihexagonal prism, {hki0}; and the hexagonal bipyramid, {$hh\bar{2}hl$} are common crystal forms for holohedral trigonal and hexagonal crystals. This underlines the close relationship between the two systems.

Rhombohedron, {$h0\bar{h}l$}. In the hexagonal system the face ($h0\bar{h}l$) results in a bipyramid. In the ditetragonal scalenohedral class, with only one set of vertical planes of symmetry, ($h0\bar{h}l$) results in a new form, a rhombohedron, which consists of six rhombic faces. It can be regarded as a cube distorted by compression along a threefold axis (an *obtuse rhombohedron*) or an extension in the same direction (an *acute rhombohedron*). A rhombohedron can be orientated in two ways, one positive, one negative (Figure 47). The two rhombohedra can be present together and

symmetry, but this is not the case here. The explanation is that a threefold axis perpendicular to a symmetry plane is equivalent to a sixfold inversion axis, and such a combination will belong to the hexagonal system. The symmetry content of the ditetragonal scalenohedral class is: three planes of symmetry intersecting in a threefold axis; three two-

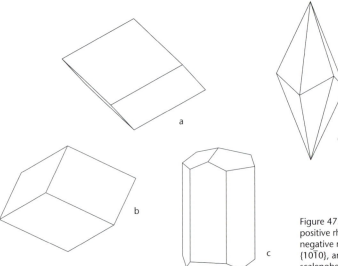

Figure 47. Trigonal crystal forms: (a) positive rhombohedron, {10$\bar{1}$1}; (b) negative rhombohedron, {01$\bar{1}$1}, c), {10$\bar{1}$0}, and {01$\bar{1}$2}; (d) ditrigonal scalenohedron, {$hkil$}.

38

Figure 48. Calcite from Derbyshire, England. Subject:16 × 17 mm.

when uniformly developed will resemble a hexagonal bipyramid.

Ditrigonal scalenohedron, {*hkil*}. The general form consists of twelve scalene triangular faces. Calcite belongs to this class.

The ditrigonal pyramidal class, 3*m*

The twofold axes and centre of symmetry are absent, and the threefold inversion axis is accordingly reduced to an ordinary threefold axis. The form {0001} is a *pedion* of a single face; {10$\bar{1}$0} is a *trigonal prism* and {*hki*0} a *ditrigonal prism*. The general form is a *ditrigonal pyramid*, {*hkil*}. This and the {0001} form show clearly that the principal axis is polar, i.e. that the two ends of the axis are symmetrically different. This is also shown by the piezo-electric and pyroelectric properties of crystals. Tourmaline belongs to this class.

The trigonal trapezohedral class, 32

In the trigonal trapezohedral class there are no planes of symmetry, only a threefold axis and three twofold axes. The twofold axes are polar, and so piezoelectricity and pyroelectricity are also exhibited in this class. The general crystal form is a *trigonal trapezohedron* of six faces, each with an irregular quadrilateral outline. For every trapezohedron there

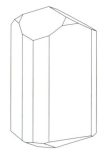

Figure 49. Tourmaline: {01$\bar{1}$0}, {11$\bar{2}$0}, {10$\bar{1}$1}, {02$\bar{2}$1}, {01$\bar{1}\bar{1}$}, and {10$\bar{1}\bar{2}$}.

39

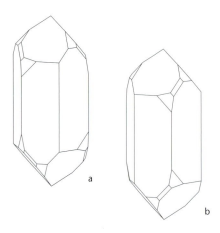

Figure 52. Quartz: (a) left-handed quartz with {10$\bar{1}$0}, {10$\bar{1}$1}, {01$\bar{1}$1}, {2$\bar{1}\bar{1}$1} and trapezohedron {6$\bar{1}\bar{5}$1}; (b) right-handed quartz with {10$\bar{1}$0}, {10$\bar{1}$1}, {01$\bar{1}$1}, {11$\bar{2}$1} and trapezohedron {51$\bar{6}$1}.

Figure 50. Tourmaline from Stack Nala, Pakistan. Field of view: 4258 mm.

Figure 53. Quartz from Brazil. Field of view: 40 × 50 mm.

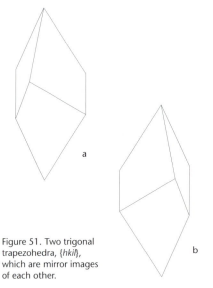

Figure 51. Two trigonal trapezohedra, {hkil}, which are mirror images of each other.

can exist another trapezohedron that is a mirror image of the first, just as the right hand is reflected by the left. Quartz is the best-known example of this class. It occurs either as right-handed or left-handed quartz.

The trigonal rhombohedral class, 3

In the trigonal rhombohedral class every face cutting the principal axis generates a *rhombohedron*, which accordingly is the general crystal form. Dioptase belongs to this class.

The orthorhombic system

The orthorhombic prism appears in cross-section as a rhomb, which has given its name to the orthorhombic system. Crystals of this system are often simply called *rhombic;* this is unfortunate, because they may then be confused with the trigonal rhombohedron. The prefix *ortho* (Gk. 'right') underlines the characteristic 90° angle between the planes of symmetry and the twofold axes. The orthorhombic system contains three classes.

The orthorhombic bipyramidal class, $2/m2/m2/m$

The orthorhombic bipyramidal class is a holohedral class in which there are three planes of symmetry that are perpendicular to each other. In consequence there are also three twofold axes perpendicular to each other and a centre of symmetry. Each twofold axis forms the intersection of two planes of symmetry and is perpendicular to the third. The complete symbol is $2/m2/m2/m$. The twofold axes are selected as reference axes and the three axes, *a*, *b*, and *c* are of un-

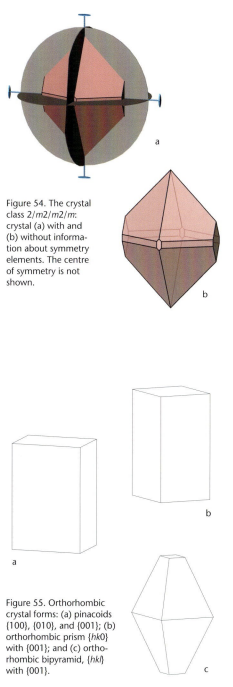

Figure 54. The crystal class $2/m2/m2/m$: crystal (a) with and (b) without information about symmetry elements. The centre of symmetry is not shown.

Figure 55. Orthorhombic crystal forms: (a) pinacoids {100}, {010}, and {001}; (b) orthorhombic prism {*hk*0} with {001}; and (c) orthorhombic bipyramid, {*hkl*} with {001}.

41

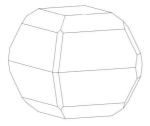

Figure 56. Sulphur: {001}, {010}, {101}, {011}, {111}, and {113}.

Figure 57. Baryte from Sardinia, Italy. Field of view: 60 × 74 mm.

equal length. In this system there is no principal axis of a higher rank, and there are thus three equally valid ways of orientating the crystal.

Crystal forms:

Pinacoid, {100}, {010}, or {001}. A face parallel with a symmetry plane is repeated once; *e.g.* {100} includes (100) and ($\bar{1}$00).

Prism, {*hk*0}, {*h*0*l*}, and {0*kl*}. A face parallel to one axis and cutting the other two is repeated three times to give a four-sided prism with a rhombic cross-section. It is of no consequence to which axis the prism is parallel.

Figure 58. Topaz from Thomas Range, Utah, USA. The crystal is seen along a twofold axis. Subject: 10 × 14 mm.

Orthorhombic bipyramid, {*hkl*}. A face cutting all three axes is repeated seven times to give an orthorhombic bipyramid, the general crystal form of the class. Many minerals belong to this class, e.g. sulphur, baryte, olivine, and topaz.

The orthorhombic pyramidal class, *mm*

In the orthorhombic pyramidal class only two planes of symmetry are present. They are (of necessity) perpendicular to each other and there is consequently a twofold axis of symmetry at their intersection. This axis, a polar axis, is always used as the vertical (*c*) axis in the orientation of the crystal. Of the special crystal forms, the pinacoids {100} and {010} are present, but the {001} form is reduced to a *pedion*. This form, perpendicular to the polar axis, consists of only one face, (001); the other face, (00$\overline{1}$), is an independent pedion. The prism {*hk*0} is preserved, but the forms {*h0l*} and {*0kl*} are reduced to domes, each with only two faces. The general form {*hkl*} is a pyramid of four faces and is an open form. Hemimorphite and natrolite belong to this class.

Figure 59. Hemimorphite: {100}, {010}, {001}, {110}, {301}, {031}, and {12$\overline{1}$}.

The orthorhombic sphenoidal class, 222

In the orthorhombic sphenoidal class the three twofold axes of symmetry are present

Figure 60. An orthorhombic sphenoid, {*hkl*}.

but the planes of symmetry are absent. This results in a general crystal form of four faces, *a sphenoid*, {*hkl*}, which can be thought of as a distorted tetrahedron. Crystals of the class are either right-handed or left-handed. Epsomite belongs to this class.

The monoclinic system

The monoclinic system includes three classes that all have one axis (*a*) inclined (Gk. *mono* = 'one', *clino* = 'incline'). (Compare the orthorhombic system, which has three axes at right angles.) In the monoclinic system $a \neq b \neq c$; $\alpha = \gamma = 90°$; and β, the angle between the *a* and *c* axes, is $\neq 90°$.

The monoclinic prismatic class, 2/m

The monoclinic prismatic class, a holohedral class, has one plane of symmetry perpendicular to a twofold axis of symmetry. There is also a centre of symmetry. The symbol for the class is 2/m. Monoclinic crystals are usually orientated with the twofold axis as the *b* axis and with the *a* and *c* axes contained in the symmetry plane. Prominent edges between crystal faces are chosen as the directions of the *a* and *c* axes; this normally gives the simplest indices.

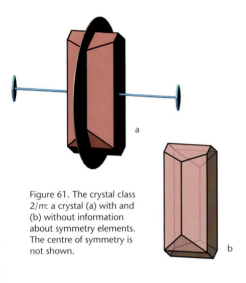

Figure 61. The crystal class 2/m: a crystal (a) with and (b) without information about symmetry elements. The centre of symmetry is not shown.

Figure 64. Orthoclase from Arendal, Norway. Field of view: 50 × 61 mm.

Figure 62. Monoclinic crystal forms: (a) {001}, {101}, and a monoclinic prism {110}; (b) orthoclase with {010}, {001}, {110}, {10$\bar{1}$}, and {20$\bar{1}$}; (c) an amphibole with {100}, {010}, {110}, {120}, and {021}.

Figure 63. Epidote from Untersulzbachtal, Austria. Subject: 28 × 72 mm.

Crystal forms:

Pinacoid, {010}, which is parallel to the plane of symmetry, and {100}, {001}, and other {h0l} forms parallel to the twofold axis of symmetry are all crystal forms composed of two faces.

Prism, {hkl}. A face in a general orientation to the twofold axis and the plane of symmetry is repeated three times to form a four-faced prism. Such a prism can be parallel to the *a* or the *c* axis and still be a general crystal form. Gypsum, epidote, some feldspars, and most pyroxenes, amphiboles, and micas belong to this class.

The monoclinic domatic class, *m*

The twofold axis is absent in the domatic class. The general form {hkl} is a *dome* consisting of one face and its 'mirror image'. Scolecite and kaolinite belong to this class.

The monoclinic sphenoidal class, 2

In the sphenoidal class the only element of symmetry is a twofold axis. The axis is a polar axis, which means that in this class we also have right-handed and left-handed crystals. The general form {hkl} is a *sphenoid*. A monoclinic sphenoid has only two faces and is an open form, as opposed to other sphenoids.

The triclinic system

The triclinic system has two classes, a holohedral and a hemihedral, and represents the lowest degree of symmetry. The name *triclinic* refers to the fact that all three axes are inclined; i.e. the angles are ≠ 90°, and the three axes *a*, *b*, and *c* are different in length. The orientation of a triclinic crystal is arbitrary. A choice of three prominent crystal edges as the directions of reference axes will give the simplest indices.

The triclinic holohedral class, $\bar{1}$

A centre of symmetry characterizes the triclinic holohedral class. A centre of symmetry can also be regarded as a onefold inversion axis, which explains the symbol of the class, $\bar{1}$. There are no special crystal forms in this class. Every face is repeated by the centre of symmetry as a corresponding face diametrically opposite to it, forming a *pinacoid*, {hkl}, the general form. Axinite, kyanite, and plagioclase feldspars are examples of this class. The triclinic hemihedral class, 1, has no symmetry element at all, not even a centre of symmetry. The symbol 1 indicates that every face will be 'repeated' after a rotation of 360°, i.e. not repeated. Thus, every crystal form {hkl} consists of only one face.

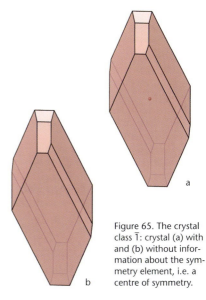

Figure 65. The crystal class $\bar{1}$: crystal (a) with and (b) without information about the symmetry element, i.e. a centre of symmetry.

Figure 66. Axinite from Bourg d'Oisans, Dauphiné, France. Field of view: 18 × 27 mm.

Formation and growth of crystals

Crystals are formed from hydrous solutions, from fluid 'magma' (rock melt), or directly from gases. In such states the chemical constituents are randomly distributed, but when conditions such as pressure, temperature, or concentration are changed the ordered crystalline state can arise.

Halite crystals are formed from a hydrous solution. Such a solution, e.g. sea water, can carry a certain quantity of Na^+ and Cl^- ions, and this quantity is governed by a number of conditions. The solution can become saturated by a further supply of ions, by evaporation of water from the solution, or by a decrease in temperature. In either case, NaCl will crystallize from the solution and release some energy by ordering the ions in a regular three-dimensional pattern. The solution can for a period be oversaturated before the process of crystallization begins, because the ions need nuclei to provide centres for crystallization. Whether one large crystal or many small crystals are formed depends on several factors. Roughly speaking, rapid crystallization favours the formation of many small crystals, whereas slow crystallization will favour the formation of fewer and larger crystals. It is essentially the availability of space that determines the degree of external perfection displayed by individual crystals.

Crystal formation from hydrous solutions takes place not only in near-surface environments such as shallow seas or salt lakes, but also from hydrothermal solutions, i.e. warm hydrous solutions rich in ions moving upwards under relatively high pressure through fissures and cavities from sources at depth in the earth's crust. Many of the most beautiful crystal specimens are formed in this way.

Crystals are also formed from melts, either

simple melts of the same chemical composition as the water–ice pair, or from more complex melts such as the magmas from which volcanic rocks are formed. In the solidification of magma, many different types of ions are present and the crystallization of several minerals becomes a complicated process, which we shall not discuss further here.

Crystal formation also takes place as a result of changes in pressure and temperature. Minerals already formed may then become unstable and be transformed into other minerals. In this solid-to-solid transformation, such as occurs in some metamorphic rocks, recrystallization takes place mainly by diffusion of ions, partly at grain boundaries, partly through crystal lattices.

Finally, crystallization can take place directly from the gaseous state, a process known as *sublimation*. It is common in volcanic regions, where crystals of sulphur and other minerals form in this way.

Crystal faces

In most crystals some faces are more developed than others. This happens because growth, like many other features of crystals, varies in accordance with direction in the crystal lattice. As mentioned above, a crystal is a three-dimensional lattice in which the lattice points determine a series of lattice planes, which are the only possible crystal faces.

During crystal growth the faces that grow most rapidly will eliminate themselves, whereas the faces that grow most slowly will become prominent. Growth takes place more readily at edges and corners than in the middle of a face, where there are initial difficulties. Rapid growth can nevertheless occasionally take place on the middle of a face. This is due to the phenomenon called *spiral growth*, in which deposition follows an edge created in the middle of the face by a dislocation in the lattice.

Figure 67. Spessartine from White Pine Co., Nevada, USA. In fissures and cavities of rocks there are good possibilities for crystal growth from hydrothermal solutions. Field of view: 18 × 27 mm.

Figure 68. 'Byssolite' with epidote from Knappen-wand, Untersulzbachtal, Austria. 'Byssolite' is a local name for fibrous actinolite or tremolite. Field of view: 40 x 45 mm.

Figure 69. Pectolite from New Jersey, USA. Pectolite commonly occurs in aggregates of acicular (needle-shaped) or fibrous crystals.
Subject: 53 × 62 mm.

Figure 70. Prehnite from Brandenberg, Namibia. Prehnite is mostly seen in botryoidal aggregates with a hackly surface. Field of view: 56 × 84 mm.

Figure 71. A two-dimensional illustration of the relationship between the density of points in a lattice and the distances between lattice planes. With some reservations it can be stated that the greater density of points, the more likely it is that a lattice plane will appear as a crystal face.

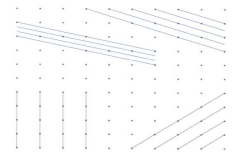

Ideal crystals are rare. Lattice defects and empty sites or foreign ions in the lattice are common, as are chemical heterogeneity (zoning) with associated changes in colour.

Crystal habit and aggregates

A complete description of the external shape of a crystal provides a combined statement of the types of crystal forms it displays (bear in mind that the term *form* has a specific meaning in crystallography) and of the way in which these crystal forms are developed in relation to each other: whether the crystal is long or short, fibrous, needle-shaped, prismatic, tabular, or equidimensional (i.e. having the same dimensions in all directions), and so forth. In combination these characteristics are expressed as the *habit* of the crystal. Paired characteristics, such as *long prismatic*, or simple terms referring to the prevalent crystal form, such as *octahedral*, are commonly used to express the habit of a crystal.

An *aggregate* is an assemblage of crystals. It is described as granular if the crystals are roughly equidimensional, and it can be fine-, medium-, or coarse-grained according to the size of the crystals. An aggregate can also be

foliated, scaly, hair-like, fibrous, radiating, columnar, or wire-like. Other terms can be used according to the character of the crystals in the aggregate. Some minerals form aggregates that are dense, massive, banded, stalactitic, shaped like a bunch of grapes (botryoidal), kidney-shaped (reniform), or form incrustations or coatings. Others are earthy or are found as small spheres (oolitic) or as somewhat larger pea-like spheres (pisolitic). Aggregates can also have moss-like forms (dentritic) or forms like branches of trees (arborescent).

In practice, the distinction between a crystal habit and the shape of an aggregate is often somewhat vague.

Figure 72. Agate from Zweibrücken, Pfalz, Germany. Agate is a banded variety of chalcedony with alternating layers of different colours, usually in concentric forms. Subject: 69 × 95 mm.

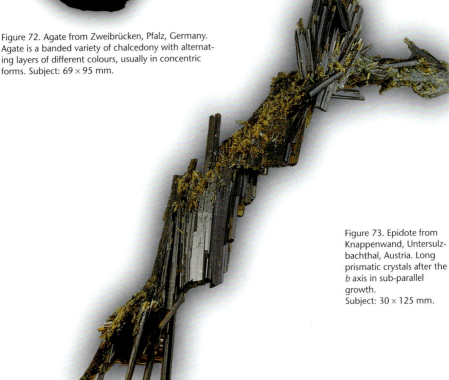

Figure 73. Epidote from Knappenwand, Untersulzbachthal, Austria. Long prismatic crystals after the *b* axis in sub-parallel growth.
Subject: 30 × 125 mm.

Figure 74. Tremolite from Kragerø, Norway. As shown here, tremolite can develop as asbestos, i.e. as ultra-thin fibres. Field of view: 29 × 49 mm.

Parallel growth

Parallel growth of several crystals of the same mineral is common and is often seen in aggregates. It can also be found in more isolated crystal groups, as in amethyst. A special kind of parallel growth occurs when crystals of two minerals with closely related structures have grown together with crystallographic directions in common. This phenomenon is called *epitaxy* and is known from kyanite–staurolite intergrowths where the two *c* axes are parallel and the (100) plane in kyanite and the (010) plane in staurolite constitute the interface. It is also known from hornblende–augite intergrowths, in which hornblende has grown epitaxially on augite.

Figure 75. Quartz var. amethyst, a sceptre crystal from Hanekleivtunnelen, Holmestrand, Vestfold, Norway. This special mode of growth is seen most often in amethyst. Subject: 23 × 45 mm.

Figure 76. Fluorite from Elmwood Mine, Carthage, Tennessee, USA. An unusual example showing solution of a crystal that is related to a crystallographic direction. A section along a threefold axis is in part resisting the solution process. Subject: 87 × 95 mm.

Pseudomorphism

A mineral can be transformed into another mineral without a corresponding change in the outline of the crystal. The result is a *pseudomorph*, formed when a change in physical and chemical conditions is such that the original mineral becomes unstable and is transformed into another mineral while preserving its original outline. Pseudomorphism can take place to various degrees, ranging from a rearrangement of the chemical components without a chemical change as, for example, in a calcite pseudomorph after aragonite, to a complete change in chemical composition, as with copper pseudomorphs after aragonite.

Etching

Crystals can be attacked by solvents, either naturally or artificially. Solution—or *etching*

as it is usually termed when referring to attack on crystal faces—has different effects with respect to different crystallographic directions.

Etch figures or *solution pits* are small depressions on crystal faces produced by the action of a solvent. As these figures have shapes that are related to the structure of the crystal, they can reveal the true symmetry of the crystal, which may not be obvious from the combination of crystal forms that is present.

Twin crystals

Two or more crystals of the same species can occur grown together in a particularly symmetrical manner. Such crystals are called *twins*, and the symmetry operation necessary to bring one crystal into a position corresponding to the other is called a *twin law*.

Some twin laws are named after minerals that display a particular type of twinning; e.g. the spinel law is named after the mineral spinel. A twin law can also be named after a well-known locality where the twinning in question has been found; e.g. the Carlsbad law typical of some feldspars, named after the spa town formerly known as Carlsbad in the Czech Republic.

There are various types of twins, and accordingly there are various ways of expressing twin laws. The laws are specified in terms of the concepts of *twin axes* and *twin planes*. A *twin axis* is a direction about which one crystal will have to be rotated 180° in order to be in the same orientation as its twin. A twin axis cannot be identical with a twofold, fourfold, or sixfold axis, since these axes already incorporate a rotation of 180°. A *twin plane* is a plane in which one crystal is reflected into its twin. The twin plane is always parallel to a possible crystal face, but by its nature it cannot be a plane of symmetry. For crystals having a centre of symmetry, twins can be described both by a rotation and by a reflection.

Twins can be *contact twins* with a composition plane, often the twin plane itself, or *penetration twins* in which one crystal appears to have grown through the other. A distinction is also made between *simple twins* and twins resulting from *repeated* or *polysynthetic* twinning, where a large number of individual crystals are in twin positions. Polysynthetic twins typically occur in thin lamellae, as in plagioclases, where they can be recognized as thin striations on {001}.

Where three or more individuals are twinned, one can speak of *trillings*, *fourlings*, etc. or simply of *multiple twins*.

Many minerals show twinning, some even in accordance with several different twin laws. Calcite, aragonite, fluorite, spinel, rutile, quartz, and feldspars are well-known examples of minerals that display twinning; they are treated in the mineral descriptions in Part II. Minerals that are commonly twinned generally have crystal structures characterized by an incomplete symmetry

Figure 77. Calcite, twinned on {0001}, on fluorite, from Elmwood Mine, Carthage, Tennessee, USA. Field of view: 54 × 81 mm.

Figure 78. Calcite, twinned on $\{01\bar{1}2\}$, from an unknown locality. Field of view: 30×30 mm.

element. This particular element is usually part of the twinning operation, as in aragonite. Aragonite is orthorhombic but pseudo-hexagonal, i.e. closely resembling hexagonal crystals. This is illustrated by the angles between the faces of the forms {110} and {010}, which are very close to 60°. If they were exactly 60° the c axis would be a sixfold axis and not just a twofold axis, and {110} would be a plane of symmetry. By forming twins with {110} as the twin plane aragonite simulates hexagonal symmetry, as shown in Figure 80. Twinning of this type is often referred to as *mimetic* twinning.

Some twins have been formed during growth; others were formed later during a transformation, i.e. a transition from one crystal structure to another; and others yet again have been formed as a result of stress caused by geological processes.

The chemical properties of crystals

Atoms and elements

An atom is the smallest unit of matter that characterizes a particular chemical element. It consists of a nucleus of protons and neutrons around which a number of electrons are rotating. Protons are positively charged, neutrons are neutral, and electrons negatively charged. The number of protons determines the species of the element and signifies its atomic number (Z). Atoms of the same element can have differing numbers of neutrons, and these different types of atom are called *isotopes*.

A neutral atom has an equal number of protons and electrons, whereas a positively or negatively charged atom has a surplus of protons or electrons respectively. Charged atoms are called *ions*; positive ions are *cations* and negative ions are *anions*.

The electrons, each having a mass of about 1/2000th that of a proton, circle round the nucleus at distances that make the atom 10,000–20,000 times larger than the nucleus. They are found in orbits or energy levels at various distances from the nucleus. Special rules govern the number of electrons in each orbit or electron shell. The innermost shell can contain a maximum of two electrons, the next shell eight, the next again eighteen, and so forth. Shells containing many electrons are further subdivided into levels; for example, the shell with a maximum of eighteen electrons is subdivided into levels of two, six, and ten (maximum).

The arrangement of the electrons is called the *electron configuration*, and this determines the chemical properties of an element. Some configurations are more stable than others. Particularly stable configurations are those in which the outermost electron shell is complete. These configurations contain the max-

Figure 79. Rutile from Magnet Cove, Arkansas, USA. The ring consists of eight individuals twinned on {101}. Subject: 15 × 18 mm.

Figure 80. Aragonite from Mingla-nilla, Cuenca, Spain. The two pseudohexagonal crystals are in random orientation with respect to each other. Each is made up of three individuals twinned on {110}. Subject: 40 × 53 mm.

55

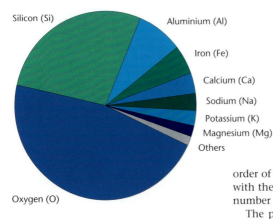

Figure 81. The most common chemical elements of the earth's crust. The pie diagram shows their relative frequency measured by weight.

order of 1 ångstrøm ($1\text{Å} = 10^{-10}$ m) but varies with the number of electrons: the higher the number of electrons, the larger the radius.

The periodic table of the elements and a list of selected elements with name, symbol, and atomic number will be found on pages 421-422.

imum number of electrons. The inert gases (helium, argon, etc.) have complete outermost shells. Other atoms readily accept or donate electrons in order to obtain a similar configuration. In doing so, they become electrically charged. Thus, sodium (Na, $Z = 11$) and chlorine (Cl, $Z = 17$) in their ground state have the electron configurations 2, 8, 1 and 2, 8, 7. By donating an electron sodium changes to Na^+ with an electron configuration of 2, 8, 0; chlorine, by accepting an electron, becomes Cl^- with the configuration 2, 8, 8. For both elements the outermost level or sublevel of electrons will then be complete.

Atoms are roughly spherical, and in most instances the crystalline state can be considered in terms of a three-dimensional packing of spheres. The radius of the spheres is in the

Chemical bonding

The forces that bind the atoms, ions, and molecules together in a crystal determine its physical and chemical properties. A distinction is traditionally made between four types of chemical bonding: the *ionic bond*, the *metallic bond*, the *covalent bond*, and the *van der Waals bond*. Transitional forms exist between these types: e.g. the bond between silicon (Si) and oxygen (O) in silicates is regarded as intermediate between an ionic bond and a covalent bond.

The ionic bond

As the name implies, the ionic bond exists between ions. Ions are electrically charged atoms that have given up electrons to become cations (positive ions) or have received electrons to become anions (negative ions). Ions of opposite charge attract each other; the greater the charge, the smaller the distance, the greater the attraction. Ultimately, however, the ions will come so close together that the attraction will turn into a repulsion.

K	Ca	Mg	Al	Si	O
	Na	Fe			

Figure 82. The most common elements in the earth's crust: oxygen (O), silicon (Si), aluminium (Al), iron (Fe), calcium (Ca), sodium (Na), potassium (K), and magnesium (Mg). They are shown as the ions O^{2-}, Si^{4+}, Al^{3+}, Fe^{2+} or Fe^{3+}, Ca^{2+}, Na^+, K^+, and Mg^{2+}, and are drawn in approximately correct proportions; for example, O^{2-} has a radius of 1.4 Å, and Si^{4+} a radius about 0.3 Å (1 Å = 10^{-10} m).

A cation will attract anions from all directions. It will thus tend to surround itself with as many anions as possible while at the same time keeping other cations at a distance. Anions will act in the same way. The fact that the bonds are effective in all directions is the main reason for the high degree of symmetry that characterizes crystals with pure ionic bonding. How many anions can surround a cation and vice versa is simply a matter of space, which again is a matter of the ratios between the sizes of the ions in question. In halite, NaCl, the ratio between Na^+ and Cl^- is 0.97 Å/1.81 Å = 0.54. This means that six Na^+ ions can be accommodated around every Cl^- ion and six Cl^- ions around every Na^+ ion. This is usually expressed by saying that the coordination number is six for both ions. In caesium chloride, CsCl, the ions are almost of equal size, which results in a coordination number of eight. Fluorite, CaF_2, has twice as many F^- as Ca^{2+} ions, and in this mineral Ca^{2+} is coordinated with eight F^-, whereas F^- is coordinated with four Ca^{2+}.

The physical and chemical properties of halite are characteristic of crystals with ionic bonding, namely, poor thermal and electrical conductivity, relatively high melting point, high solubility, lack of colour or only pale colour, brittleness and good cleavage, relatively low density, and moderate hardness. Properties such as hardness and melting point are, however, highly dependent on the strength of the ionic bond, which again is largely determined by the charge (valence) of the ions involved. For example, halite, NaCl, has a hardness of 2½, whereas periclase, MgO, with the same structure, has a hardness of 6 because of the greater strength of the bonds between the divalent Mg^{2+} and O^{2-} ions. The ionic bond is the most common bond in minerals.

The metallic bond

Metals are elements that readily give up their outermost electrons. The result is the formation of positive ions, cations that are held to-

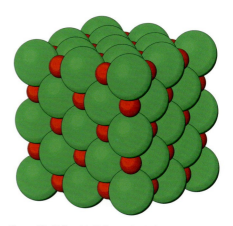

Figure 83. Halite, NaCl, has an ionic bond. Every Cl^- (green) is coordinated with (i.e. surrounded by) six Na^+ (red), and every Na^+ is coordinated with six Cl^-. The bond between an ion and its surroundings is directed spherically in all directions and not in any particular direction.

gether as densely packed spheres by the detached electrons. The electrons are said to form a 'cloud of electrons' that can move freely between the cations. This is the explanation for the high thermal and electrical conductivities that are typical of metals. Metals are also characterized by poor hardness and by being very malleable (i.e. easy to shape by hammering, etc.). These properties are due to the fact that the strength of the bond is moderate and the bonds are not constrained to certain directions but extend in all directions. The electrons are effective in absorbing light, and this causes the metals to be opaque (i.e. impenetrable to light) and to have a high lustre. As the crystal structures of metals take the form of dense packing of equal-sized spheres with the bonds in all directions, they have high symmetries and are often cubic.

The three-dimensional packing of spheres can be regarded as a sequence of layers of closely packed spheres in which the spheres of one layer fit into the hollows in the layers above and below. Each layer is consequently displaced in relation to its neighbouring layers. This displacement can result in two different arrangements: one cubic, the other

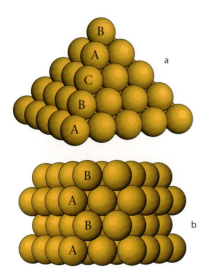

Figure 84. (a) Cubic closest packing of spheres, with a sequence of layers ABCABC; (b) hexagonal closest packing of spheres, with a sequence ABABAB.

hexagonal. In both cases a sphere will have twelve nearest neighbours. In the *cubic closest packing* of spheres the sequence of layers is ABCABC, i.e. every third layer is in an identical position (Figure 101). In the *hexagonal closest packing* of spheres the sequence of layers is ABABAB, in which every second layer is in an identical position.

Gold, silver, and copper are well-known minerals with the metallic bond and having cubic close packing of spheres.

The van der Waals bond

The van der Waals bond is a weak bond that is common in organic compounds and in crystals of inert gases such as neon and argon. Few minerals have this type of bond. The best known are sulphur, in which eightfold rings of sulphur atoms are held together by van der Waals bonds, and graphite, in which van der Waals bonds hold the carbon layers together. As would be expected, compounds with only van der Waals bonds have low melting points and poor hardness.

The covalent bond

An atom is most stable when the outermost electron shell is complete, as in the inert gases. Chlorine has seven electrons in the outermost level, and one more is required to attain the stable condition with eight. As mentioned above, chlorine readily accepts an electron to form an ionic bond, and thus becomes negatively charged. There is, however, another way for a chlorine atom to form a compound, which is to share an electron with another chlorine atom as if they both have eight electrons in the outermost level. This type of bond, the *covalent bond*, is formed within the boundaries of the individual atoms and is directed towards the partnering atom. Covalent compounds are very stable and they generally have low solubility, high melting points, poor conductivities, and great hardness.

The best-known example is diamond, which is pure carbon (C). The carbon atom has four electrons in the outermost level. By sharing electrons with four neighbouring atoms, each carbon atom will have its outermost level completed by eight electrons. In this way every carbon atom will be strongly bonded to four neighbours in specific direc-

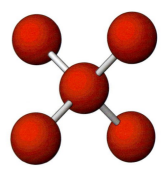

Figure 85. In diamond, C, every carbon atom shares electrons with four neighbours. In consequence, all atoms have eight electrons in their outermost shell. This covalent bonding is a very strongly directional and accounts for the unique hardness of diamond.

Figure 86. Olivine from Lanzarote, Spain. Olivine encompasses a complete solid-solution series between the end-members forsterite and fayalite. Subject: 40×60 mm.

tions. This bonding confers the extreme hardness that is characteristic of diamond.

Few other minerals have pure covalent bonding, but many minerals, such as the silicates, have a type of bond that is intermediate between a covalent bond and an ionic bond.

Polymorphism, isomorphism, and polytypism

Some chemical compounds can exist as minerals with different crystal structures. Such minerals, known as *polymorphs*, are formed under different pressure and temperature conditions. Diamond and graphite, marcasite and pyrite, calcite and aragonite, and quartz, tridymite, and cristobalite are well-known examples. The transition from one polymorph to another commonly requires chemical bonds to be broken, which can result in a slow and often incomplete process. Other transitions between polymorphs take place momentarily and reversibly, as in high and low quartz, in which the transition is a mat-

ter of minor adjustments in the lattice without any breaking of bonds.

A number of minerals have the same basic crystal structure and a related—but different—chemical composition. In consequence, they have several properties such as crystal symmetry and cleavage in common. Such minerals are called *isomorphs*. Minerals of the calcite group (calcite, siderite, smithsonite, and rhodochrosite) are good examples. This group is, however, different from other isomorphic groups in not forming true solid-solution series. Such series are much better displayed in other mineral groups such as the garnets, in which elements readily substitute for each other as long as they fit into various positions in the crystal structure. Where the charges on the ions replacing each other are different, a balance can be achieved by a paired substitution, as in feldspars (where the substitution of Si^{4+} by Al^{3+} is linked to a substitution of K^+ by Ca^{2+}).

Some minerals form complete isomorphic solid-solution series, as in the olivines, which are mixtures of the two end-members forsterite, Mg_2SiO_4, and fayalite, Fe_2SiO_4. When a solid-solution series is complete, the end-members can be mixed in all proportions. In some solid-solution series the miscibility is limited to certain compositional ranges. This occurs in the plagioclase series at low temperature, where the members are not completely homogeneous but actually consist of lamellae with slightly different compositions.

Some minerals are so closely related that the only distinction between them is a difference in the stacking of parts of the structure. Such minerals are called *polytypes*. They are particularly common in layer silicates such as the micas. They consist of sequences of uniform layers that can be stacked in different ways, almost like the layers in the cubic and hexagonal closest packing of spheres.

The physical properties of crystals

Hardness

The hardness of a mineral is defined as its resistance to being scratched. Hardness is a characteristic property because it expresses the degree of coherence between the chemical constituents of the crystal. The hardness of a mineral is determined primarily by the type of chemical bonding, and secondarily by the size and valency of the constituent atoms or ions.

The classical Mohs scale is used to specify the hardness of a mineral. It was introduced in 1822 and is still very useful. Mohs scale makes use of common minerals and is based on the principle that a mineral of hardness 4 can scratch one of hardness 3, but not one of hardness 5, and so on. The intervals of the scale vary but this is of no practical importance. Minerals with hardness 1–2 can be scratched with a fingernail, those of hardness 3–5 with a knife; hardness 6 is equivalent to the hardness of a knife, and minerals of hardness 7–10 can scratch a knife. This traditional comparison with a knife works with old-fashioned knives of non-stainless steel, but modern stainless knives are harder. The state-

ment that a mineral has a hardness of $2\frac{1}{2}$ means only that the hardness is somewhere between 2 and 3. The Mohs scale is as follows:

Talc	Feldspar
Gypsum	Quartz
Calcite	Topaz
Fluorite	Corundum
Apatite	Diamond

Minerals with metallic bonds such as gold, silver, and copper are relatively soft, with hardness about 3. Minerals with covalent bonds are very hard because the bonding is within the outermost electron shells of the atoms, as in diamond, with hardness 10. The hardness of minerals with ionic bonding is primarily determined by the size and valency of the ions involved and therefore varies considerably; halite and fluorite, both with ionic bonding, have hardnesses of $2\frac{1}{2}$ and 4 respectively. Minerals with van der Waals bonds have poor hardness; for instance, sulphur has hardness 2.

Since the hardness is largely determined by the crystal structure, is it a property that varies according to the crystallographic direction in which it is measured. There is in general no notable variation in hardness with direction in the crystal, but in kyanite the difference is significant. Kyanite usually forms long prismatic crystals. Along these crystals the hardness is $4\frac{1}{2}$; across them it is $6\frac{1}{2}$.

In practice the hardness of a mineral is tested by trying to scratch an unaltered surface by means of a fingernail, a knife, or another mineral of known hardness. Hardness can also be estimated by scratching the mineral in question with materials of known hardness, such as window glass (H about 5). In either case one should, of course, exercise great care not to damage good crystal faces or crystal terminations, but should limit the test to less important areas. One should also make sure that the test is directed to the mineral itself and not to an aggregate, where one might merely test the coherence of individual grains.

During hardness tests other physical properties can also be observed. This applies in

Figure 87. Kyanite from St. Gotthard, Switzerland. Kyanite commonly forms long prismatic crystals, which have a hardness of 4½ along the length of the crystals and a hardness of 6½ across their length. Field of view: 60 × 90 mm.

particular to properties such as brittleness, sectility, malleability, ductility, and toughness. These properties can provide useful information and are generally easy to estimate with a knife or hammer. For instance, it is characteristic of gold to be extremely malleable (to be hammered into very thin sheets) and ductile (to be drawn into a wire), for pyrite to be brittle (easily broken), for acanthite to be sectile (easily cut into flakes with a knife), and for jadeite to be tough (strong and flexible). Some minerals such as talc are flexible (easily bent); others, such as most micas, are both flexible and elastic, returning to their original shape after being bent.

Cleavage and fracture

The cleavage of a crystal is defined as its ability to break along planes whose directions, like those of crystal faces, are determined by the internal structure of the crystal. The crystal can have one or several cleavage planes or faces that belong to one or more crystallographic forms. According to the readiness with which a crystal will break along these planes one can speak, e.g., of *perfect, good, distinct, indistinct, poor,* or *no cleavage.*

A cleavage face usually differs from a crystal face by being smoother and more reflecting. It will also often have a pearly lustre because air has intruded cleavage planes below the surface. A crystal does not have to be completely broken in order to evaluate the cleavage properties: cleavage planes are quite often faintly visible within the crystal as signs of incipient cleavage. If the specimen is expendable, it can be hammered into pieces, whose fresh faces can be examined.

As cleavage is a property determined by the crystal structure it is also governed by the symmetry elements of the structure. A complete record of the cleavage of a mineral therefore has to include information about

61

Figure 88. Halite from Hallein, Salzburg, Austria. Halite has a cleavage parallel to the cube faces, i.e. in three directions perpendicular to each other. Field of view: 100 × 150 mm.

the directions of cleavage and the angles between these directions as well as an estimate of the perfection of the cleavage.

Cubic crystals can have cleavage parallel to the cube faces, i.e. in three directions perpendicular to each other (galena, halite), parallel to the octahedron, i.e. in four directions (fluorite), or to the rhombic dodecahedron, i.e. in six directions (sphalerite). If a specimen has only one direction of cleavage, it cannot be cubic but must be a crystal with pinacoidal cleavage and therefore tetragonal (apophyllite), trigonal or hexagonal (molybdenite), orthorhombic (topaz), monoclinic, or triclinic (mica). If a specimen has two cleavage directions it cannot be cubic, trigonal, or hexagonal but, for instance, monoclinic and having cleavage parallel to two pinacoids (feldspar) or to a prism that

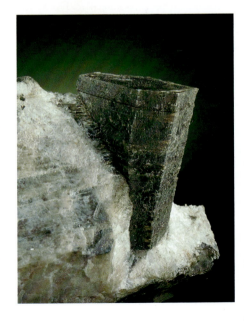

Figure 89. Muscovite from Bamble, Norway. Like all micas, muscovite has a perfect cleavage parallel to one plane. Field of view: 60 × 77 mm.

Figure 90. Hornblende from Arendal, Norway. Hornblende is an amphibole and has a cleavages parallel to the {110} prism faces, with angles of 124° and 56° between the two cleavage directions. Subject: 80 × 107 mm.

also includes two directions (pyroxenes and amphiboles). In pyroxenes and amphiboles, both with prismatic cleavage parallel to the {110} prism, the angles between the cleavage directions are diagnostic: in pyroxenes the angles between the cleavage planes are 93° and 87°, whereas in amphiboles they are 124° and 56° (Figures 455c and 468c).

A crystal with three cleavage directions can be cubic, as stated above, but if all three directions are not perpendicular to each other the crystal might be an orthorhombic mineral such as baryte with a combination of pinacoidal and a prismatic cleavage. If all three directions of cleavage are oblique to each other, the crystal could have rhombohedral cleavage and be calcite. Other minerals with the calcite structure, such as siderite, have the same type of cleavage. This shows that the cleavage is determined primarily by the type of crystal structure and not by the chemical composition of the mineral.

Some minerals have no cleavage but break with irregular surfaces. Such fracture can be *conchoidal*, as in quartz, where it resembles the outside of a bivalve shell, *splintery*, *hackly* (i.e. jagged with sharp edges), or *uneven*, without any pattern. Other minerals, such as

Figure 91. Rutile needles in quartz from Novo Horizonte, Bahia, Brazil. Quartz has a density of 2.65, which is within the typical range for many light rock-forming minerals. In rutile the atoms are particularly densely packed, which results in a high density of 4.2. Subject: 73 × 113 mm.

gold, do not readily break but can be hammered into extremely thin plates.

Parting is not a proper cleavage but a tendency to break along twin planes, as seen in corundum.

Density

The density of a mineral is defined as its mass per unit volume. It is usually expressed in g/cm³. The density depends on the type of atoms composing the mineral and how densely they are packed in the crystal lattice.

Most minerals have densities that fall within two ranges: (1) around 2.7, common for many of the light-coloured rock-forming minerals such as the feldspars; and (2) around 6, common for pyrite and other ore minerals. Some important rock-forming minerals, such as the pyroxenes and amphiboles, however, have densities around 3.3.

The density of a mineral can be assessed by weighing it in the hand. One quickly learns to judge the density, and to estimate whether it lies outside the main ranges. A light-coloured mineral such as baryte with a density of 4.5 is clearly sensed as heavier than common light-coloured rock-forming minerals, and even small deviations can be noticed with practice. Density is measured by various methods based on Archimedes' principle or by suspension in heavy liquids of known density. It can also be calculated when the chemical composition and the unit cell dimensions are known.

Many mineralogy textbooks give the specific gravity of a mineral rather than its density. The specific gravity of a mineral is the ratio between its weight and the weight of an equal volume of water at 4 °C. The values for specific gravity and density are almost identical. In this book density is preferred.

The optical properties of crystals

Visible light is only a small part of the electromagnetic spectrum. It ranges from violet to red light, i.e. with wavelengths within an interval of about 4000 to 7500 Å. Light of only one wavelength, *monochromatic* light, is of only one colour, whereas a blend of the whole spectrum results in white light.

When light strikes a crystal face, some is reflected and some enters the crystal. The light entering the crystal has to conform to the internal structure of the crystal. It cannot be transmitted within the crystal with the same velocity as in a vacuum because of the impediments it meets in the form of atoms and their electrons. The ratio between the velocity of light in a vacuum and in a material is called the *refractive index* of the material; the greater the reduction in velocity, the higher the index of refraction. Values of refractive indices are greater than 1; e.g., quartz has a refractive index of about 1.55. In the general case, the reduction in the velocity of light entering a crystal is associated with a change in direction of the light rays.

In glasses and cubic crystals the refractive index is independent of the direction in which the light ray travels. Such materials are called *optically isotropic*. In other crystals, i.e. all non-cubic crystals, the refractive indices are dependent on the directions of propagation and vibration of the light. These crystals are called *optically anisotropic*. In such crystals, the atoms and their electrons are ordered in such a way that the velocity of light will vary with the direction within the crystal. In consequence, the crystal will have a range of refractive indices varying from a lowest to a highest value. The difference between the two extreme values is the *bire-fringence* of the crystal. Furthermore, in anisotropic crystals a light ray cannot vibrate at random in all directions but is split into two rays that are confined to vibrate in only two planes perpendicular to each other. Light that is confined to vibrate in one plane only is said to be *polarized*.

Part of the light entering a crystal is absorbed and the remainder leaves the crystal. How much light is absorbed depends to some extent on the thickness of the crystal, but in particular on its chemical constituents and the type of bonding between them. These factors also determine which parts of the spectrum are absorbed and which parts pass through. If the amount of light transmitted through the crystal is relatively large, the crystal is described as *transparent*. If the amount of light transmitted is moderate, the crystal is *translucent*; and if the amount is small or close to zero the crystal is *opaque*.

Figure 92. Gold from Brush Creek Mine, Sierra County, California, USA. Gold is one of the heaviest minerals; pure gold has a density of 19.3. Field of view: 26 × 39 mm.

Lustre

The appearance of a crystal face in reflected light is known as its *lustre*. The intensity of the lustre is largely determined by the amount of light that is reflected; it commonly increases with increase in the refractive index. The colour, on the other hand, has very little influence on lustre. There are two main types of lustre: *metallic* and *non-metallic*. A few minerals have *submetallic* lustre. Metallic lustre is a characteristic property of opaque minerals and is typical of metals, sulphides, and some oxides.

Non-metallic lustre includes a number of types. *Vitreous lustre* is a glass-like lustre that is very common among rock-forming minerals such as feldspars with refractive indices between 1.4 and 1.9. *Adamantine lustre* is found in diamond and other minerals whose

Figure 93. Pyrite with quartz from Peru. Pyrite has a metallic lustre, whereas quartz has a vitreous lustre. Subject: 24 × 30 mm.

refractive indices are greater than 1.9, e.g. sphalerite. In addition, one can speak of *greasy lustre*, *pearly lustre*, and *silky lustre*, which are self-explanatory terms for various surface conditions.

Colour

There are two main causes of colour in minerals. One is connected with refraction and the scattering of light as it passes through a mineral; the other—and more important—is related to the absorption of parts of the incident light, which affects the wavelengths of the light leaving the mineral.

Absorption colours
When white light containing all the colours of the spectrum passes through a mineral unchanged, the mineral appears to be white or colourless. If all parts of the spectrum are equally absorbed, the resulting colour will be grey to black according to the degree of absorption. If, however, certain wavelengths are selectively absorbed, the mineral will display a colour that is determined by the wavelengths that pass through it.

The absorbed light represents energy that has been used to move electrons from one energy level to another. The specific wavelengths that are required to do this vary from element to element and are determined by the electron configuration of the element in question.

Transition of electrons from one level to another is especially common in elements of the first group of transition elements in the periodic table. This group includes titanium (Ti), vanadium (V), chromium (Cr), manganese (Mn), iron (Fe), cobalt (Co), nickel (Ni), and copper (Cu). These elements, often referred to as *chromophores*, have the special characteristic that the outermost electron shell is not completed before a new level further out is occupied. This results in considerable mobility among the outer electrons. The energy required to move them within the same atom or from atom to atom is of the

order of the energy of visible light. Exactly which wavelengths will be used (and absorbed) depends upon several factors, including the type of chromophore, its valence state, the type of bonding to neighbour atoms, the numbers of neighbouring atoms, and features of local symmetry in the surroundings of the chromophore.

Absorption of light is often very complicated, particularly when several chromophores or the same chromophore in different valence states, or both, are present. The colours derived from absorption are therefore best illustrated by some examples.

The chemical composition of olivine is a mixture of two end-members, forsterite, Mg_2SiO_4, and fayalite, Fe_2SiO_4. Pure forsterite is colourless, whereas fayalite is dark bottle-green. This is caused by the Fe^{2+} ion absorbing more of the reddish and the violet parts of the spectrum. Common olivine is thus more or less green according to its iron content.

The garnets are a series of silicates with considerable variation in chemical composition and accordingly also in colour. Almandine, $Fe_3Al_2Si_3O_{12}$, has the same Fe^{2+} ion that in olivine gives a green colour, but in almandine it results in a reddish colour. The difference is associated with the number of neighbouring oxygen atoms, which is eight in almandine and six in olivine.

Azurite and malachite have characteristic colours, blue and green respectively. They are both hydrous copper carbonates, but the Cu^{2+} ion is placed in different environments, resulting in a difference in absorption. The red ruby and the green emerald both owe their colour to the Cr^{3+} chromophore. Ruby is a red corundum, Al_2O_3, whereas emerald is a grass-green beryl, $Al_2Be_3Si_6O_{18}$. In both minerals Cr^{3+} is present in small amounts, replacing Al and surrounded by six oxygen atoms. The difference is believed to lie in the chemical bond, which in corundum is considered to be a regular ionic bond whereas in a silicate such as beryl it is a partly covalent bond.

Alexandrite, a gemstone variety of chrysoberyl, $BeAl_2O_4$, is also coloured by small

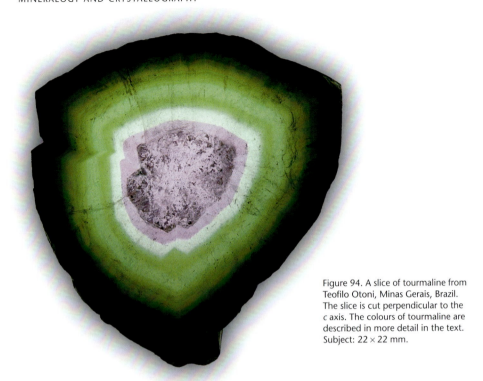

Figure 94. A slice of tourmaline from Teofilo Otoni, Minas Gerais, Brazil. The slice is cut perpendicular to the *c* axis. The colours of tourmaline are described in more detail in the text. Subject: 22 × 22 mm.

amounts of Cr^{3+}. It has an absorption spectrum that is intermediate between those of ruby and emerald. In consequence, it appears reddish or greenish according to the distribution of energy in various forms of white light. In ordinary daylight alexandrite is greenish; in 'warmer' artificial light it becomes reddish.

In the examples given so far the colour has been caused by electron transitions within single atoms, interacting with the surrounding atoms. Electron transitions from atom to atom can also give rise to colour, whether they are between atoms of the same type or between atoms of different types. The valency of individual ions then becomes blurred, and some of the 'valence electrons' become the common property of several ions. An important example of this type of 'charge transfer' is found between Fe^{2+} and Fe^{3+} in such minerals as micas, amphiboles, pyroxenes, and tourmalines.

In tourmalines the colour varies, not only from crystal to crystal, but also within the same crystal, in several ways. The ends of the crystal can be green while the middle of the crystal is red, or the crystal can be divided into a series of colour zones parallel to the principal axis. The possible variations of colour in tourmaline and their causes are manifold. The simplest examples are: pink (Mn^{3+}), yellow-brown (Fe^{3+}), and various shades of green (Fe^{2+}, Cr^{3+}, or V^{3+}).

In tourmaline and other minerals such as cordierite the phenomenon of *pleochroism* is seen. This is a variation in colour in relation to the direction in the crystal. As mentioned above, all crystals except cubic crystals are anisotropic; i.e. their optical properties, including absorption, vary with the direction within the crystal lattice. For pleochroic minerals the variation is such that it affects the colours of the mineral.

Vivianite, $Fe_3(PO_4)_2 \cdot 8H_2O$, is an example of a colour change caused by a change in the valency of an element, in this case Fe. When mined, vivianite is almost colourless, but shortly afterwards it becomes dark bluish-green because some of the iron is oxidized by the air and changes from Fe^{2+} to Fe^{3+}.

Magnetite, $Fe^{2+}Fe^{3+}_2O_4$, is a classic example of a mineral that is black and opaque because the entire visible spectrum is absorbed owing to continuous electron transitions between Fe^{2+} and Fe^{3+} which blur the valence state of these ions.

Blue sapphire, another variety of corundum, is believed to be blue because of the presence of small amounts of Fe and Ti which replace Al. The two elements are present as $Fe^{2+} + Ti^{4+}$ or $Fe^{3+} + Ti^{3+}$. Transitions between these two combinations result in absorption of the red end of the spectrum, giving the well-known cornflower-blue colour.

Colour is also imparted by defects in the crystal structure, which can result in electron transitions. In fluorite, vacancies can be pro-

Figure 95. Fluorite from Illinois, USA. The origin of the violet colour of fluorite is described in the text. Subject: 70×98 mm.

Figure 96. Quartz var. amethyst from Veracruz-Llave, Mexico. The origin of the colour in amethyst is described in the text. Field of view: 48×41 mm.

69

Figure 97. Azurite from Chessy, Lyon, France. Azurite is blue and malachite is green. Both minerals owe their colour to the same chromophore, Cu^{2+}. The chromophore, is, however, in different positions in the crystal lattices of the two minerals and absorbs different wavelengths from the visible light. Subject: 78×110 mm.

duced during growth at positions that would otherwise be sites for F^- ions. These positions are instead occupied by extremely mobile electrons that absorb all except the violet end of the visible spectrum.

Another type of structural defect that gives rise to colour is found in varieties of quartz, SiO_2. In smoky quartz small amounts of Si^{4+} are replaced by Al^{3+} and corresponding amounts of other ions, typically Na^+ or H^+, which balance the charge. Gamma or X-ray radiation from the surroundings can repel one electron of a pair in the Al–O bond, which is transferred to another site in the structure. The remaining electron continually shifts between the two positions now available, and this results in an absorption that produces a brownish smoky colour. If the crystal is heated to about 400°C, the captured electron can return and the smoky colour then disappears. Amethyst, another variety of quartz, owes its violet colour to the same principle. In this instance, however, an Fe^{3+} ion is the initial source of the colour.

Other colour phenomena
Colours can be caused by impurities. Halite containing inclusions of oil drops can be brown; quartz can be coloured by inclusions of chlorite or hematite.

Gold and other metals have metallic bonding in which every atom has contributed one or more electrons to a shared 'electron cloud'. These free electrons absorb all incident light and quickly return most of it, giving the characteristic high metallic lustre. The composition of the transmitted light can vary for very different reasons, even within the same mineral. Their colour is therefore not always a reliable characteristic for such metals, but the *streak*—the colour produced as a streak on unglazed porcelain—is far more characteristic.

Some crystals show a special 'play' of colour, which is due to various optical phenomena such as refraction or reflection. In the feldspar labradorite, the colours are caused by repeated refraction and reflection in a series of exsolutions or twin lamellae.

The famous play of colour in diamond, its 'fire', is due to *dispersion*, in which white light is split into colours because of significant differences between the refractive indexes for the shorter and the longer wavelengths of the light.

In precious opals a particular play of colour, *iridescence*, is seen. This is an interference phenomenon. These opals are built up of regular lattices of sub-microscopic spheres, and the dimensions of these lattices make them suitable to function as optical gratings. The lattices of common opals are less regular, and they have a milky appearance.

Streak

The streak of a mineral is its colour when powdered. It is investigated by rubbing the mineral on a plate of unglazed porcelain, or, if the mineral is as soft as graphite, on a piece of paper. For many minerals the streak is a more reliable property than the colour, because the true colour may be hidden by a tarnished surface. Although the streak is not a characteristic property of silicates and most other light minerals, it is a useful characteristic of sulphides and some oxides.

Luminescence

Some crystals emit light when exposed to a physical influence. This phenomenon, called *luminescence*, is normally weak and can be seen only in the dark. (It is quite distinct from incandescence.) A distinction is made between *triboluminescence*, induced by pressure or crushing; *thermoluminescence*, induced by heating; and *fluorescence* and *phosphorescence*, which are induced by ultraviolet radiation.

Radioactivity

Uranium (U) and thorium (Th) are radioactive elements that decay spontaneously, emitting α-, β-, or γ-rays. Minerals containing U and Th are affected by this radiation.

The isotopes U^{238}, U^{235}, and Th^{232} disintegrate to the end-product lead at a definite rate regardless of their mode of occurrence or pressure–temperature conditions. If the rate of decay and the amounts of the isotopes involved are known, the time of crystallization can be calculated. This relationship has been used extensively in determining the ages of geological materials.

Minerals with even a very low content of U or Th can have their crystal lattices destroyed by the radiation emitted from these radioactive elements. This process of destruction is known as *metamictization*, and the affected minerals are said to be *metamict*. The result of the process is a non-crystalline solid (a glass) with a greater volume, which commonly produces fissures in the surrounding rock. Minerals commonly seen in a metamict state include zircon, allanite, and monazite. Smoky quartz and other minerals owe their colour to radioactivity from nearby minerals.

Figure 98. Tourmaline from Grotta d'Oggi, Elba, Italy. Tourmaline is strongly pyroelectric. Field of view: 33×63 mm.

Piezoelectricity and pyroelectricity

In crystals without a centre of symmetry some directions in the crystal lattice are polar, i.e. points on the opposing sides of such axes are not symmetrically related. Examples are the threefold axis in tourmaline and the two-

Figure 99. Magnetite from the Doshkesan mine, Azerbaijan. Magnetite is among the few minerals that are strongly attracted to a magnet. Subject: 26 x 32 mm.

fold axes in quartz. The polarity is demonstrated by an electric potential between the ends or poles when the crystal is affected by heat (*pyroelectricity*) or pressure (*piezoelectricity*). Detection of piezo- or pyroelectric properties can be the only means of allocating a crystal to a particular crystal class.

Pyroelectricity in tourmaline can be demonstrated by warming a long crystal and holding it close to cigar ash, which will then jump on to the end of the crystal.

Piezoelectricity in quartz is of considerable technological importance. Slices of quartz of appropriate thickness and crystallographic orientation are used for controlling frequencies in radio components, watches, etc.

Magnetism

Magnetite and pyrrhotite are among the very few minerals that are strongly attracted by a magnet. On occasion, these minerals are found as natural magnets (lodestones), and they were at one time used for navigation. All minerals are influenced by a magnetic field, but generally only to a small degree. *Diamagnetic* minerals are repelled by a magnetic field, whereas *paramagnetic* minerals are attracted; *ferromagnetic* minerals are heavily attracted. The fact that minerals are influenced to varying degrees by a magnetic field is used in mining to separate ore minerals in the crude state.

MINERAL DESCRIPTIONS

Systematics based on genetic relations, as we know it in zoology and botany, is not possible in mineralogy. The classification of minerals is based first on chemical composition and secondly on the crystal structure, in a sequence beginning with those that are chemically more simple.

Minerals are described here in the traditional sequence, beginning with the elements and proceeding to sulphides, halides, oxides, carbonates, sulphates, phosphates, and silicates. A few small groups of minerals, such as nitrates and vanadates, are treated under the groups they most resemble. The silicates—the largest and most important of all groups—are further subdivided according to the structural linkage of the basic silicate unit, the silicon–oxygen tetrahedron. A brief section on organic minerals concludes the systematic descriptions.

There is no unambiguously and internationally accepted sequence within the individual groups, and small differences are seen from book to book, especially between the major English and German textbooks. One significant distinction between two comprehensive textbooks, the German Klockmanns *Lehrbuch der Mineralogie* (1978) and the American *Dana's New Mineralogy* (1997), is the position of quartz and the other SiO_2 minerals, which in the German book are classified as oxides and in the American book as silicates. The latter scheme is used here.

The space given here to a mineral is governed by its significance for the keen amateur mineral collector rather than the professional mineralogist. For example, the very rare mineral boléite is included because of its attractive crystal form and beautiful colour, which are highly appreciated by collectors. The scientifically very important clay minerals are, on the other hand, mentioned only briefly, since they are in general unattractive to the private collector; they normally occur as microscopic particles and are in practice indistinguishable from each other without special laboratory equipment.

As a supplement to the systematic descriptions, some of the most common minerals are listed in tables at the end of the book. The lists are ordered according to easily determinable properties, and reference is made to the more comprehensive descriptions in Part II.

Densities are given throughout the mineral descriptions and tables in grams per cubic centimetre.

Figure 100. Gold from Colorado, USA. Subject: 51 × 59 mm.

Native elements

Few chemical elements occur as native elements in the earth's crust, none of them in large quantities. Some of these minerals, e.g. gold (Au), silver (Ag), and diamond (C), are nevertheless among the best known, for man has searched for them from earliest times because of their valuable properties.

The native elements are divided into metals, semi-metals, and non-metals. The metals are composed of spherical cations like Au^+ in close packings, most of them in cubic closest packing. The cations are held together by a relatively weak chemical bond that extends in all directions. This structure is responsible for the most characteristic properties of the metals, i.e. high thermal and electrical conductivity, metallic lustre, poor hardness, high ductility and malleability, absence of cleavage, and high density. The semi-metals and non-metals have different structural properties, which are mentioned under the individual minerals.

Gold, Au

Crystallography. Cubic $4/m\bar{3}2/m$; crystals uncommon, mostly as {111}, sometimes in combinations with {110} or {100}; occurs mostly in leafy, scaly, wire-like, or dendritic forms; also massive as nuggets; twins on {111} common.

Physical properties. No cleavage, hackly fracture; very malleable and ductile, can be hammered out to extremely thin threads or plates; sectile. Hardness 2½–3; density 19.3. Pure gold is yellow; a minor content of Ag gives a lighter colour, whereas presence of Cu creates a reddish tint. Streak is shiny yellow; metallic lustre; opaque.

Chemical properties, etc. A complete solid-solution series exists between Au and Ag, and native gold almost always contains some Ag, typically 2–20%. Cu and other metals are normally present only in small quantities. The crystal structure of gold is based on a cubic closest packing of Au^+ cations.

Names and varieties. Gold with more than 20% Ag is called *electrum*.

Occurrence. Gold is a rare element in the earth's crust. Most gold occurs as native gold, while compounds of gold and Te, Bi, Sb, and Se are very rare. Gold is typically found in hydrothermal quartz veins and dykes related to Si-rich igneous rocks, commonly in association with pyrite and other sulphides. Occurrences of this type are found in California and other states in the western USA. Gold is also found finely dispersed in major ore deposits, as in Boliden, Sweden, where it is associated with pyrite and arsenopyrite.

Gold is also found as a secondary mineral as placer gold in sediments derived from weathering of primary gold deposits. Because of its great resistance to physical and chemical weathering, gold survives transport and settles in river beds, where because of its high density it is concentrated in certain layers. In such deposits gold is present both as minor grains from the primary occurrences and as major irregular lumps or 'gold nuggets', some of which are precipitated from solutions. The largest deposits of this type occur in the Republic of South Africa, for instance the fa-

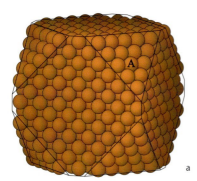 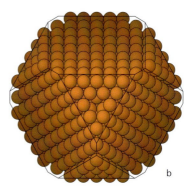

Figure 101. Cubic closest packing of spheres as found in gold, silver, and copper, illustrated here by a crystal with cubic and octahedral faces. The two crystals in (a) and (b) are identical except in two respects: first, the crystal on the right is orientated with an octahedral face parallel to the plane of the page; secondly, the ion A in the crystal on the left is omitted in (b) so that the three neighbouring atoms in the layer below can be seen. These three neighbouring atoms and the six neighbouring atoms in the layer of A, together with the three neighbouring atoms normally situated in the layer above, together make a total of twelve neighbouring atoms. On any crystal face some neighbouring atoms will inevitably be lacking.

Figure 102. Gold on quartz from Michigan Bluff Mine, Placer County, California, USA. Subject: 50 × 80 mm.

mous Witwatsrand conglomerate, and various places in Siberia.

Use. Au has long been used as a monetary standard, and considerable amounts of Au are locked up in bullion. Au bullion, medals, etc. are important objects of investment. Au is also used in jewellery, in dentistry, and for components in instruments. In order to increase its hardness Au is generally alloyed with other metals such as Ag and Cu. The amount of Au in an alloy was in the past stated in carats, i.e. parts per 24. Today it is given as per mille (‰).

Diagnostic features. Shiny yellow colour and streak. Gold is distinguished from minerals such as pyrite and chalcopyrite with similar colours by its inferior hardness, higher malleability, and density.

Figure 103. Gold from Roşia Montană (formerly Verespatak), Romania.
Subject: 14×19 mm.

Silver, Ag

Crystallography. Cubic $4/m\bar{3}2/m$; well-developed crystals rare; more common are arborescent or wire-like networks; most common are massive, tabular, or scaly forms; good crystals have simple crystal forms such as {100} or {111}; twinning on {111} is seen.

Physical properties. No cleavage, hackly fracture, very malleable. Hardness 2½–3; density 10.5. Colour is silver-white on a fresh surface, but tarnishes to grey or black; streak silver-white; metallic lustre; opaque.

Chemical properties, etc. A complete solid-solution series exists between Au and Ag. Silver commonly contains small amounts of Au or Cu, more rarely Hg, As, or Sb. The crystal structure of silver is based on a cubic closest packing of Ag^+ cations.

Figure 104. Silver from Kongsberg, Norway. Subject: 160×260 mm.

Figure 105. Silver, twinned on {111}, from Kongsberg, Norway. Subject: 43 × 56 mm.

Names and varieties. *Amalgam* is a natural alloy of Ag and Hg.

Occurrence. In small amounts silver is widespread in the upper oxidized zones of ore deposits. Large concentrations of silver are found in veins and dykes precipitated from hydrothermal solutions. In such occurrences silver can be associated with calcite, quartz, fluorite, zeolites, and various sulphides, as in Kongsberg, Norway; it is associated with arsenides and sulphides of cobalt and nickel or uraninite, as in Freiberg, Germany, and Jachymov (formerly Joachimsthal), Czech Republic. Silver also occurs together with gold in gold–quartz veins in various places in the western USA and with copper in the famous copper deposit on the Keweenaw Peninsula, Michigan, USA.

Use. As an Ag ore native silver is of minor importance; more important are sulphides of Ag. Ag is used in photographic emulsions, electrical components, jewellery, cutlery, tableware, and was in the past used for coins.

Diagnostic features. High malleability, high density, streak, and, in part, colour on fresh surfaces.

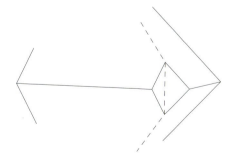

Figure 106. Silver, twinned on {111} as in Figure 105. The dotted lines mark the twin plane (111). All faces on both individuals belong to the {100} form.

81

Copper, Cu

Crystallography. Cubic $4/m\bar{3}2/m$; good crystals rare, most often distorted; crystal forms are mainly {111}, {100}, {110}, and {hk0}; occurs generally in irregular masses or in dentritic or wire-like forms; twinning on {111} is seen.

Physical properties. No cleavage, hackly fracture, very malleable. Hardness 2½–3; density 8.9. Colour is copper-red on fresh surfaces, but readily tarnishes to dark brownish colours; streak copper-red, shiny; metallic lustre; opaque.

Chemical properties etc. Native copper often contains small amounts of Ag, As, Sb, Bi, or Hg. The crystal structure of copper is based on a cubic closest packing of Cu^+ cations.

Occurrence. The most important occurrences of native copper are associated with basaltic lavas in which copper is formed as a result of reactions between hydrothermal solutions and Fe-rich minerals. The largest occurrence of this type is found on the Keweenaw Peninsula, Michigan, USA. Here basaltic lavas alternate with layers of sandstone and conglomerates, and cavities in these formations are filled with copper in association with, e.g. calcite, epidote, other copper minerals, zeolites, and small amounts of silver. The single masses can be large; one lump weighed more than 500 tonnes. Native copper in minor amounts is also found in the oxidized zones of Cu ores together with cuprite, malachite, and azurite.

Use. Native copper is of minor importance as a Cu ore compared with Cu sulphides. Cu is primarily used for electrical purposes, in particular for cables, etc. It is also used in many alloys, such as brass (with Zn) and bronze (with Sn).

Diagnostic features. Colour on fresh surfaces, streak, malleability, and density; is often associated with malachite.

Figure 107. Copper with malachite (green) from Corocoro, La Paz, Bolivia. Subject: 95×113 mm.

Figure 108. Copper, pseudo-morphous after aragonite, from Caracoles, Antofagasta, Chile. Subject: 18 × 16 mm.

Platinum, Pt

Crystallography. Cubic $4/m\bar{3}2/m$; crystals rare; mostly found as grains or nuggets.

Physical properties. No cleavage, hackly fracture, malleable. Hardness 4–4½, high for a native metal; density 21.5, but generally lower owing to presence of other metals. Colour is steel-grey to dark grey; streak whitish-grey, bright; metallic lustre; opaque; sometimes magnetic because of Fe content.

Chemical properties, etc. Native platinum usually contains some Fe and minor amounts of elements such as Pd, Rh, Os, Ir, or Cu. The crystal structure is based on a cubic closest packing of Pt^+ cations.

Names and varieties. The name 'platinum' is derived from the Spanish *plata*, silver, owing to its similarity to silver.

Occurrence. Native platinum occurs as dispersed grains in dunite and other ultrabasic rocks, usually associated with olivine, chromite, and magnetite. It is also of secondary occurrence as grains or nuggets in sediments derived from the weathering of Pt-containing dunite rocks. The most important occurrences of platinum are in the Urals area in Russia and in the Bushveld complex in the Republic of South Africa.

Use. Pt is used as catalysts in vehicle exhausts and for similar purposes in the oil and chemical industries. Its high melting point, great hardness, and resistance to chemicals make it useful in many other industrial applications.

Diagnostic features. Colour, streak, and density; unlike silver, does not tarnish.

Figure 109. Platinum nugget from Suchowi-simsk, Urals, Russia. Weight 91 g. Subject: 41 × 20 mm.

Iron, Fe

Crystallography. Cubic $4/m\bar{3}2/m$; crystals not seen. Terrestrial iron is found as grains or nodules; meteoritic iron as masses (see below).

Physical properties. Cleavage poor on {100}, hackly fracture, malleable. Hardness 4½; density 7–8. Colour is grey to black; streak grey, bright; metallic lustre; opaque. Strongly magnetic.

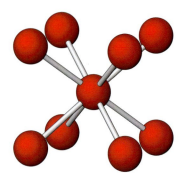

Figure 110. The crystal structure of iron is a cubic packing of ions, in which every ion has eight closest neighbours. The ions in the drawing are reduced in size in relation to the distances between them in order to reveal the structure.

Chemical properties, etc. Iron contains variable amounts of Ni and commonly small amounts of elements such as Co, Cu, or Mn. The crystal structure of iron is based on a cubic close packing in which each cation has eight nearest neighbours.

Names and varieties. The mineral iron is also called α-Fe.

Occurrence. Iron, which is believed to constitute the bulk of the earth's core, is a very rare mineral at the earth's surface, where it is unstable and is readily oxidized. Iron formed on the earth is called terrestrial iron. The best-known occurrence of terrestrial iron is at Uiffaq (formerly Ovifaq) on Qeqertarsuaq (Disko Island), Greenland, where the iron occurs in basalts, ranging from tiny grains up to masses of several tonnes. The iron was formed by reduction of Fe oxides when the basalt intruded carbonaceous layers.

The element Fe (but not the mineral iron) is the main constituent of iron meteorites and is also present in stony meteorites. The iron-bearing minerals *kamacite* and *taenite* found in meteorites are Fe–Ni alloys; both are regarded as separate mineral species. Kamacite contains about 4–7.5% Ni; taenite contains higher percentages of Ni. Meteoritic Fe normally occurs as regular intergrowths of kamacite and taenite formed by solid-state exsolution during cooling of the parent body. The pattern of the intergrowths, known as the *Widmanstätten structure*, can be made visible by etching a polished surface.

Use. Native iron has no importance as an iron ore.

Diagnostic features. Magnetic properties and malleability; often coated with iron oxides (rust-coloured).

Lead, Pb

Cubic; crystals known in simple forms like {100} and {111}, but usually found as irregular grains or plates. Hardness 1½; density 11.4. Colour is lead-grey, tarnishing readily to almost black. Lead is a rare mineral formed under extreme reductive conditions. It is known, e.g. from the Mn-rich iron ores at Pajsberg and Långban, Sweden, and at Franklin, New Jersey, USA.

Mercury, Hg

Liquid above −39 °C; below this temperature it solidifies as trigonal crystals. Mercury is silver-white with a distinct metallic lustre. It is a rare mineral, primarily found as drops associated with cinnabar, HgS. *Amalgam* is a natural alloy of Ag and Hg.

Figure 111. Iron meteorite found in Carbo, Sonora, Mexico. Widmanstätten figures are visible on the surface, which has been polished and subsequently etched. These figures are alternating lamellae of kamacite and taenite in a crystallographically ordered pattern. The structures were formed by exsolution of two iron phases during the cooling of the core of the parent body. The two inclusions, one drop-shaped, the other circular, are troilite, FeS. Field of view: 80 × 120 mm.

Arsenic, As

Crystallography. Trigonal, $\bar{3}2/m$; crystals rare; found mostly as scaly, tuberous, or layered masses, sometimes in botryoidal or stalactitic forms.

Physical properties. Perfect cleavage on {0001}; brittle. Hardness 3½; density 5.7. Colour is tin-white on fresh surfaces but readily tarnishes to grey or black; streak grey; metallic lustre; opaque.

Chemical properties, etc. Together with Sb and Bi, As forms a group of semi-metals, whose properties are intermediate between those of the true metals and the non-metals. In the crystal structure of the semi-metals every atom is linked to three close neighbour atoms by strong bonds and to three more distant ones by weaker bonds. This results in puckered sheets with weak bonding between them, which explains the perfect cleavage. Arsenic can contain small amounts of other metals or semi-metals. The mineral is poisonous; when heated it sublimes to a gas with a distinct smell of garlic.

Figure 113. Arsenic from St. Andreasberg, Harz Mountains, Germany. Subject: 78 × 99 mm.

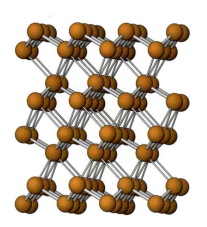

Figure 112. In the crystal structure of arsenic, antimony, and bismuth every atom is linked to three close neighbour atoms by strong bonds and to three more distant ones by weaker bonds. This creates puckered sheets that are easily cleaved from each other. The crystal structure is here viewed approximately along the sheets.

Names and varieties. *Allemontite* is a natural alloy of As and Sb.

Occurrence. Arsenic is a rare mineral typically found in hydrothermal veins associated with Ag, Co, and Ni ores, as at St. Andreasberg, Harz Mountains, Germany.

Use. Arsenic has a number of applications, e.g. in herbicides and insecticides. The most important sources of As for technical purposes are As-containing sulphides, e.g. arsenopyrite.

Diagnostic features. Mode of occurrence, brittleness, and hardness. The garlic-smelling gases arising from heating are very characteristic—and poisonous.

Antimony, Sb

Crystallography. Trigonal, $\bar{3}2/m$; crystals rare; found mostly as granular or lamellar masses, rarely in radiated or botryoidal forms.

Physical properties. Perfect cleavage on {0001}; brittle. Hardness 3–3½; density 6.7. Colour is tin-white to grey; streak lead-grey and shining; metallic lustre; opaque.

Chemical properties, etc. Antimony belongs to the semi-metal group (see arsenic). Miscible with As; can contain small amounts of Ag and Fe.

Names and varieties. *Allemontite* is a natural alloy of Sb and As.

Occurrence. Antimony is rare. It occurs in hydrothermal veins, typically together with Ag ores and stibnite, Sb_2S_3.

Use. Sb expands during solidification, and this property has been used in various applications, e.g. as type metal. Most Sb comes from stibnite and Sb-containing lead ores.

Diagnostic features. Antimony resembles the other semi-metals but usually has a coating of white Sb_2O_3.

Bismuth, Bi

Crystallography. Trigonal, $\overline{3}2/m$; crystals rare; found mostly in lamellar or massive forms; also foliated, reticulated, or arborescent.

Physical properties. Perfect cleavage on {0001}; brittle and sectile. Hardness 2–2½; density 9.7. Colour is silver-white with a reddish tinge; sometimes mottled and tarnished; streak silver-white; metallic lustre; opaque.

Figure 114. Allemontite from Challanches, Isère, France. Field of view: 60 × 69 mm.

Chemical properties, etc. Bi belongs to the semi-metals (see arsenic).

Occurrence. Bismuth is a rare mineral mostly found in hydrothermal veins, commonly in association with minerals containing Ag, Co, Ni, or Sn; also found in pegmatites.

Use. Bi has a low melting point and for that reason is used in many alloys. It is also used for medical purposes. Most of this Bi is a by-product of the mining of Sn and Pb ores.

Diagnostic features. Relatively high density, low hardness, and a reddish tinge distinguish bismuth from other semi-metals.

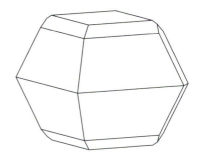

Figure 116. Sulphur: {001}, {011}, {111}, and {113}.

Sulphur, S

Crystallography. Orthorhombic $2/m2/m2/m$; well-developed crystals common, often bipyramidal; also found as massive lumps, incrustations, or coatings; sometimes stalactitic.

Physical properties. No distinct cleavage; conchoidal or uneven fracture; brittle. Hardness 1½–2½; density 2.1. Colour is sulphur-yellow but can vary in yellowish, greenish, or brownish nuances owing to impurities; streak white; greasy to adamantine lustre; transparent or translucent. Sulphur is a poor conductor of heat and therefore feels warm; it is easily melted, inflammable, and burns to sulphur dioxide.

Chemical properties, etc. Sulphur belongs to the non-metals. Its crystal structure consists of ring-shaped S_8-molecules linked together by weak molecular forces. Sulphur can contain small amounts of Se.

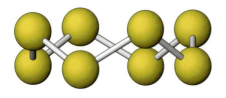

Figure 115. A single molecule of sulphur consisting of eight atoms united by strong bonds. In an orthorhombic crystal structure, such molecules are in addition bonded by weak molecular forces.

Occurrence. Sulphur is found in volcanic regions, where it is precipitated directly from S-rich gases (sublimation) or is formed by incomplete oxidation of hydrogen sulphide in gases. It can also be formed from sulphates by bacterial activity. The largest deposits of sulphur are found in evaporites such as halite, anhydrite, and gypsum. To a smaller extent sulphur is found as a weathering product of sulphides.

Use. S has many applications, e.g. as a raw material for manufacturing sulphuric acid, insecticides, and fertilizers; for vulcanizing rubber; in the oil and paper industries, etc. About half the S used for these purposes comes from mining of the mineral sulphur. The rest is a by-product from the mining of various sulphide ores.

Diagnostic features. Colour, inflammability, brittleness, and low hardness.

Diamond, C

Crystallography. Cubic $4/m\overline{3}2/m$; crystals common, often {111} or {110}, often with curved faces; twinning on {111} occurs.

Physical properties. Perfect cleavage on {111}, brittle. Hardness 10, thus the hardest mineral; density 3.5. Colourless to yellowish, also brownish or greyish, more rarely pink, red, green, or blue; can be black owing to inclusions; adamantine lustre; transparent to translucent. Very high refractive index and a

Figure 117. Sulphur from Sicily, Italy. Field of view: 90 × 150 mm.

strong dispersion are the optical properties that cause the famous 'fire' (sparkle), which is enhanced by various cuts in gem diamonds.

Chemical properties, etc. Diamond consists of pure C and is thus a polymorph of graphite. Graphite is the stable phase under low pressure–temperature conditions, diamond under high pressure–temperature conditions. This pair of polymorphs represents the larg-

Figure 118. (a) In the crystal structure of diamond every atom is bonded to four neighbouring atoms by a strong covalent bond; the C–C distance, 1.54 Å, is the same throughout. (b) In graphite every atom is strongly bonded to the three closest neighbouring atoms within the same layer (at a distance of 1.42 Å) and weakly bonded to a fourth neighbouring atom in the layer next to it (at a distance of 3.36 Å). (The latter bond is not shown in the figure.) In order to reveal the structure, the atoms are shown smaller in relation to the interatomic distances.

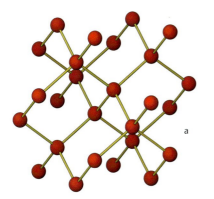

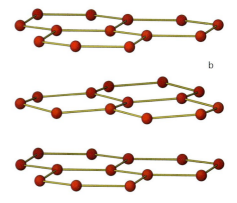

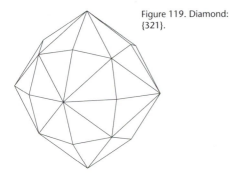

Figure 119. Diamond: {321}.

est contrast in mineralogy with respect to crystal structures and physical properties. In diamond every C atom is tetrahedrally bonded to four other atoms with a covalent bond. In consequence, all the atoms have their outermost electron shells completed, and this provides the structural explanation for the extreme hardness of diamond.

Names and varieties. *Bort* is industrial-grade diamond used for abrasives, saw-blades, etc.; *carbonado* is a cryptocrystalline (very fine-grained) variety of diamond; bort and carbonado are usually black or grey; bort can be of other colours. Neither is cut as a gemstone.

Occurrence. Diamond occurs as scattered crystals in kimberlites, which are ultrabasic rocks having their origin in the upper parts of the earth's mantle. At this depth the pressure–temperature conditions are suitable for diamonds to crystallize. The best-known deposit of this type is found at Kimberley in the Republic of South Africa. Diamond is also found in secondary deposits in sediments formed by the erosion of primary diamond-bearing rocks. In these deposits diamonds have survived the processes of weathering and transport owing to their great hardness and chemical resistance. The diamonds are usually concentrated in certain beds because of their relatively high density. Most natural diamonds come from such occurrences.

Figure 120. Diamond in kimberlite from Colesberg Kopje Mine, Republic of South Africa. Field of view: 12 × 14 mm.

Figure 121. Diamonds in gravel from an unknown locality. Field of view: 80 × 120 mm.

Figure 122. Diamond, {110}, with slightly curved faces, from the Republic of South Africa. The crystal is 8 mm across.

Figure 123. Graphite from Aasiaat, Greenland. Subject: 83 × 120 mm.

Use. Diamond is used industrially for cutting, grinding, and polishing. About 75% of the diamonds for these purposes are now synthetically produced. Diamond is among the most sought-after gemstones and is evaluated by four c's: colour, clarity, cut, and carat. (One carat equals 0.2 g.)

Diagnostic features. Diamond is recognized by its extreme hardness, high lustre, and perfect {111} cleavage.

Graphite, C

Crystallography. Hexagonal $6/m2/m2/m$; crystals uncommon, mostly simple tabular crystals dominated by {0001}; occurs most commonly as foliated, scaly, or granular masses.

Physical properties. Perfect cleavage on {0001}; cleavage folia flexible but inelastic. Hardness 1; feels greasy; density 2.2. Colour and streak black; metallic lustre; opaque.

Chemical properties, etc. Graphite consists of pure C and is thus a polymorph of dia-

mond. Graphite is the stable polymorph at low pressure–temperature conditions. It has a crystal structure in which every atom is strongly bonded to three neighbour atoms in the same sheet, whereas the bond to the fourth neighbour atom in the next sheet is a weak bond. This explains the perfect cleavage on {0001}.

Occurrence. Graphite is a common mineral in metamorphosed coal deposits, limestones, schists, and gneisses; locally it can constitute essential parts of these rocks. Graphite is to a smaller extent also found in igneous rocks and their associated pegmatites and hydrothermal veins.

Use. The high melting point (3000 °C) and chemical resistance of graphite make it useful in connection with casting and melting processes. It is also used as a lubricant, as a constituent of certain paints, and as the 'lead' in pencils.

Diagnostic features. Graphite is recognized by its hardness, cleavage, foliation, greasy character, colour, and streak (easily seen on paper). Graphite is distinguished from molybdenite by its totally black streak.

Sulphides

Figure 124. Pyrite from
Quiruvilca, La Libertad, Peru.
Subject: 73 × 104 mm.

The sulphides are a large group of minerals and include the majority of the economically important ore minerals. Usually included in the group are the corresponding but much rarer compounds in which sulphur (S) is replaced by Se, Te, As, Sb, or Bi, and also the compounds known as the sulphosalts.

A sulphide is basically an oxygen-free compound of sulphur (S) and one or more metals. In the simplest sulphides the crystal structure can be regarded as a spherical packing of large S atoms with smaller metal atoms in some of the interstices. The chemical bonding is in varying degrees a mixture of metallic, ionic, and covalent bonds.

Most sulphides have a metallic character with a strong colour and streak and metallic lustre. Most are also opaque. They have high densities and most of them, in contrast to the pure metals, are brittle.

Acanthite, Ag_2S

Crystallography. Monoclinic $2/m$; crystals uncommon, mostly in simple cubic forms replacing the cubic polymorph *argentite*, which is the stable phase above 179 °C. Crystals typically show parallel orientation in groups. Most commonly massive, thread-like, or arborescent, or as coatings.

Physical properties. No distinct cleavage;

Figure 125. Acanthite from Kongsberg, Norway. Subject: 47×49 mm.

Figure 126. Acanthite from Rayas, Guanajuato, Mexico. Subject: 25 × 36 mm.

fine conchoidal fracture; slightly flexible and very sectile. Hardness 2–2½; density 7.3. Colour is black; streak black and shining; metallic lustre, bright on fresh surfaces, otherwise dull; opaque.

Occurrence. Acanthite occurs in hydrothermal veins, commonly in association with other Ag ore minerals such as pyrargyrite, proustite, and native silver. It is also found as small inclusions in galena, making galena an important Ag ore. Freiberg in Saxony, Germany, and the Comstock lode, Nevada, USA, are among the localities for acanthite.

Use. Acanthite is an important Ag ore.

Diagnostic features. Pronounced sectility and lack of good cleavage, as compared with galena; and (to some extent) colour.

Chalcocite, Cu_2S

Crystallography. Monoclinic $2/m$; crystals uncommon, mostly as small tabular pseudo-hexagonal crystals; twinning simulating orthorhombic or hexagonal symmetry common; occurs mostly massive, or occasionally as thin coatings on other ores.

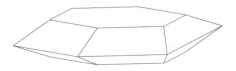

Figure 127. Chalcocite: {001}, {021}, and {113}.

Figure 128. Chalcocite from Flambeau, Ladysmith, Wisconsin, USA. Subject: 75 × 87 mm.

Physical properties. Cleavage indistinct on {110}; conchoidal fracture; more-or-less brittle. Hardness 2½–3; density about 5.7. Colour is lead-grey on fresh surfaces, but tarnishes to dull black; streak grey to black; metallic lustre; opaque.

Names and varieties. *Digenite*, Cu_9S_5, is a closely related mineral; it is bluish and is found together with chalcocite.

Occurrence. Chalcocite is especially found as a supergene mineral formed in the enriched zone of Cu-rich ore deposits in arid regions. In the uppermost zones of such deposits the primary Cu minerals are exposed to oxidizing surface waters and the Cu-rich solutions derived from them react at lower level with the primary ore, which becomes enriched by redeposition of Cu minerals such as chalcocite. Ore deposits of this type are found at Rio Tinto, Spain, Bisbee, Arizona, and several other places in the USA. Chalcocite is also found as a primary mineral in hydrothermal veins associated with, e.g. bornite, chalcopyrite, and pyrite, as at Butte, Montana, USA.

Use. Chalcocite is among the most important Cu ores.

Diagnostic features. Colour, hardness, partial brittleness, and association with other copper minerals.

Bornite, Cu_5FeS_4

Crystallography. Tetragonal $\bar{4}2m$, above 228 °C cubic; crystals rare, mostly in simple pseudo-cubic forms; occurs mostly granular, massive.

Physical properties. No distinct cleavage; conchoidal to uneven fracture. Hardness 3;

density 5.1. Colour is bronze-yellow or golden brown on fresh surfaces, but readily tarnishes to mottled bluish-red colours, streak greyish-black; metallic lustre; opaque.

Chemical properties, etc. The Cu/Fe ratio can vary widely.

Occurrence. Bornite is widespread. It occurs in association with other sulphides in hydrothermal veins, with chalcocite in the enriched zone of Cu deposits (see chalcocite), in contact-metamorphic rocks, and as scattered grains in basic rocks. It alters to other Cu minerals such as chalcocite, chalcopyrite, malachite, and azurite.

Use. Bornite is a Cu ore of minor importance.

Diagnostic features. The fresh colour and the tarnished colours.

Galena, PbS

Crystallography. Cubic $4/m\bar{3}2/m$; crystals common, most frequent crystal form is {100}, sometimes in combination with {111} or more rarely {110}; twinning on {111} seen, also on {114}, which can result in diagonal striations on cleavage surfaces; occurs also as cleavable masses, coarse to fine granular.

Physical properties. Perfect cleavage on {100}. Hardness 2½; density 7.6. Colour and streak are lead-coloured; metallic lustre, shining to dull; opaque.

Chemical properties, etc. Galena itself is normally rather pure; Ag and other elements are typically present as small inclusions of various sulphides. Galena has the same crys-

Figure 129. Bornite in quartz from Harvey Hill Mine, St-Pierre-de-Broughton, Megantic County, Quebec. Field of view: 42 × 63 mm.

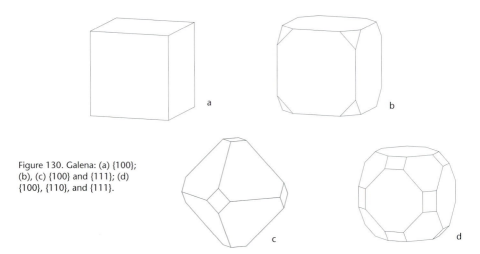

Figure 130. Galena: (a) {100};
(b), (c) {100} and {111}; (d)
{100}, {110}, and {111}.

tal structure as halite, NaCl, but with Pb instead of Na, and S instead of Cl. Accordingly Pb is coordinated with six S, and S in the same way with six Pb. While halite has a pure ionic bond, the bonding in galena is partly metallic.

Names and varieties. *Alabandite*, MnS, *altaite*, PbTe, and *clausthalite*, PbSe, belong to the galena group, but are much more rare.

Occurrence. Galena is a very common mineral and the most important Pb mineral. It is

Figure 132. Galena in quartz from Maarmorilik, Greenland. Field of view: 63×86 mm.

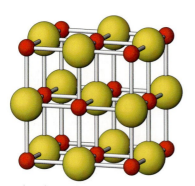

Figure 131. The crystal structure of galena can be derived from the halite structure by replacing Na with Pb (red), and Cl with S (yellow). Pb is thus coordinated with six S, and S is coordinated with six Pb.

Figure 133. Galena with siderite on quartz from Neudorf, Sachsen-Anhalt, Germany. Field of view: 120 × 180 mm.

Figure 134. Galena from Joplin, Missouri, USA. Field of view: 40 × 60 mm.

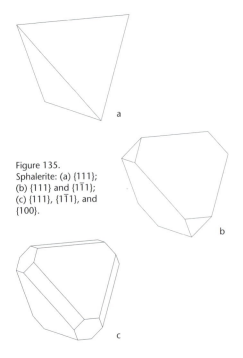

Figure 135.
Sphalerite: (a) {111};
(b) {111} and {1$\bar{1}$1};
(c) {111}, {1$\bar{1}$1}, and
{100}.

land. In Maarmorilik, Greenland, galena is found in quartz veins in association with sphalerite.

Use. Galena is by far the most important Pb ore and, as mentioned above, is also important as an Ag ore. Pb has many applications, e.g. for electric batteries, pipes, cables, ammunition, lead glass, radiation protection, various low-melting alloys and, especially in the past, for paints and as an additive in petrol.

Diagnostic features. Cleavage, density, hardness, and, to some extent, colour and streak.

Sphalerite, ZnS

Crystallography. Cubic $\bar{4}3m$; crystals common, mostly tetrahedral and commonly as a positive tetrahedron {111} in combination with a negative tetrahedron {1$\bar{1}$1}, {100}, {110}, or {hkk}; crystals frequently complex; twinning on [111] common, as simple twins, repeated contact twins, or penetration twins; occurs also in cleavable masses, coarse to fine granular, massive, rarely fibrous or botryoidal.

found in hydrothermal veins in association with other common sulphides such as sphalerite, pyrite, and chalcopyrite, and with quartz, baryte, fluorite, calcite, and many other minerals. In some deposits galena is associated with Ag ores and commonly contains inclusions of Ag sulphides in large enough amounts to make galena itself an important Ag ore. It occurs also in veins and cavities in limestones, frequently in association with sphalerite as, for instance, in what is known as the Tri-State District including parts of Missouri, Kansas, and Oklahoma, USA. This district has supplied an astonishing number of exquisite crystal groups. Galena is also found in pegmatites and contact-metamorphic rocks, and is finely dispersed in many sediments. Well-known localities for galena in Europe include Freiberg and various places in the Harz Mountains, Germany, Pribram in the Czech Republic, and Derbyshire, Cumbria, and County Durham in Eng-

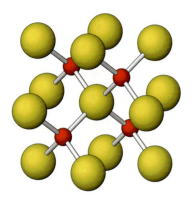

Figure 136. The crystal structure of sphalerite can be derived from the diamond structure by replacing half the carbon atoms with Zn (red) and the other half with S (yellow). Zn is thus coordinated with four S, and S with four Zn.

Figure 137. Sphalerite from Binntal, Switzerland. Field of view: 22 × 33 mm.

Physical properties. Cleavage perfect on {110}, i.e. six directions. Hardness 3½–4; density about 4.0. Colour very variable; pure ZnS is mostly yellow or brown, rarely red, green, white, or colourless, darkening almost to black with increasing content of Fe, Mn, or Cd; streak yellow to brown, darkens with increasing Fe content but usually lighter than the mineral colour; lustre varies from sub-metallic adamantine on cleavage surfaces to a somewhat greasy metallic lustre in fine-grained aggregates; transparent to opaque.

Chemical properties, etc. Zn is almost always partly replaced by Fe. The degree of Fe substitution depends partly on the amount of Fe available and partly on the temperature at the time of crystallization (the higher the temperature, the more Fe). Where sphalerite has crystallized in association with an Fe-rich mineral such as pyrrhotite, which indicates that sufficient Fe is available, the Fe/Zn ratio can be used to calculate the temperature at the time of formation. In this way sphalerite

can serve as a geological thermometer. Mn and Cd as well as Fe can replace Zn, but in general only in small amounts.

The crystal structure of sphalerite can be derived from that of diamond by exchanging half the C atoms in diamond for Zn and the other half for S; accordingly, Zn is coordinated with four S, and S with four Zn. The structure can be regarded as a cubic closest packing of S atoms with Zn in the spaces that are tetrahedrally coordinated with S. *Wurtzite*, ZnS, is a polymorph of sphalerite in which the S atoms are packed in a hexagonal closest packing of spheres. Wurtzite is a relatively rare mineral with physical properties similar to those of sphalerite.

Names and varieties. *Sphalerite* is derived from a Greek word meaning 'deceptive'; the old name *blende* has a similar meaning. Both probably refer to difficulty in identifying the mineral or to its uselessness as an ore in comparison with its regular companion galena. *Greenockite*, CdS, is a rare mineral related to wurtzite.

Occurrence. Sphalerite is a very common mineral; it occurs in much the same manner as galena and is very commonly associated with it; a number of occurrences are mentioned under galena. In addition, sphalerite is found in hydrothermal veins associated with magnetite, pyrite, pyrrhotite, and other sulphides. Transparent crystals of sphalerite are known from Lengenbach quarry, Binntal in Switzerland, and from Picos de Europa, Santander in Spain.

Use. Sphalerite is by far the most important Zn ore. Zn has many applications, e.g. in electroplating, in brass and other alloys, as sheet zinc, in batteries, pressure-creosoting, paints, and medicines. Sphalerite is also the chief source of Cd and several other rare elements.

Diagnostic features. As mentioned above, sphalerite can be difficult to identify; the most useful characteristics are cleavage, lustre, and, to some extent, hardness and mode of occurrence.

Figure 138. Sphalerite, a cleaved sample, from Chivera Mine, Sonora, Mexico. Subject: 26 × 43 mm.

Figure 139. Sphalerite on dolomite from Trepca, Serbia. Field of view: 66 × 99 mm.

Figure 140. Sphalerite (red) with galena from Joplin, Missouri, USA. Field of view: 40 × 60 mm.

Figure 141. Wurtzite from Pribram, Bohemia, Czech Republic. Subject: 53 × 61 mm.

Figure 143. Chalcopyrite from Perkiomenville, Pennsylvania, USA. Subject: 80 × 78 mm.

Chalcopyrite, $CuFeS_2$

Crystallography. Tetragonal $\overline{4}2/m$; crystals are typically dominated by a sphenoid, e.g. {112} resembling a cubic tetrahedron; combinations are seen of a positive and a negative

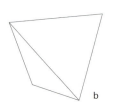

Figure 142. Chalcopyrite: (a) {111}; (b) {112}.

a　　　　　　b

sphenoid in which one form has dull or striated faces and the other has bright faces without striation; twinning on {112} is seen; occurs mostly massive.

Physical properties. No distinct cleavage; uneven fracture; brittle. Hardness 3½–4; density about 4.2. Brass-coloured, can tarnish weakly; streak greenish-black; metallic lustre; opaque.

Chemical properties, etc. The chemical composition is normally close to the ideal formula. The crystal structure can be derived from that of sphalerite by replacing half the Zn with Cu and the other half of the Zn with Fe. This results in a doubling of the c axis and a reduction of the symmetry from cubic to tetragonal.

Occurrence. Chalcopyrite is the most widespread Cu mineral and is among the most important Cu ores. It occurs finely scattered

in most igneous rocks and is locally concentrated in large masses, together with pyrite, magnetite, and pentlandite. It also occurs in hydrothermal veins, as fillings or larger lens-shaped bodies, sometimes in contact-metamorphic limestones. It is found in some sediments and is known, e.g. from the famous fossil-bearing copper slate at Mansfeld in Germany. Chalcopyrite is present in many deposits, e.g. Rio Tinto in Spain, Bisbee, Arizona, and French Creek, Pennsylvania, USA,,
Use. Chalcopyrite is one of the main Cu ores. Cu, an important metal, is used for electrical wiring, water pipes and fittings, bronze, brass, and other alloys, roofing materials, etc.
Diagnostic features. Crystal habit, colour, and streak; hardness less than pyrite and greater than gold; brittle in comparison with gold.

Stannite, Cu_2FeSnS_4

Tetragonal and structurally related to chalcopyrite; found mostly as granular masses. No distinct cleavage; hardness 4, density about 4.4. Colour is steel-grey to black; streak black; metallic lustre. Stannite occurs in hydrothermal Sn-rich veins and is locally of some importance as an ore, e.g. in Cornwall in England and in Bolivia.

Pyrrhotite, $Fe_{1-x}S$

Crystallography. Hexagonal $6/m2/m2/m$, but monoclinic, $2/m$ below about 250 °C; crystals uncommon, normally tabular parallel to {0001}; mostly massive, granular.

Figure 144. Chalcopyrite from Bottino mine, Seravezza, Italy. Field of view: 54 x 81 mm.

Physical properties. Cleavage usually not distinct, uneven fracture, brittle. Hardness about 4; density 4.6. Colour is bronze-yellow or brown; streak greyish-black; metallic lustre; opaque. Magnetic, varying in intensity.

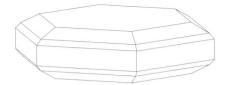

Figure 145. Pyrrhotite: {10$\bar{1}$0}, {0001}, {10$\bar{1}$1}, and {10$\bar{1}$2}.

Chemical properties, etc. There is normally a deficit of Fe in relation to S; x in the formula is typically between 0 and 0.2. The structure is a hexagonal closest packing of spheres of S atoms, in which Fe is located in the positions octahedrally coordinated with S; about one in every eight Fe positions is empty.

Names and varieties. *Troilite*, FeS, is a closely related mineral found in many meteorites as well as terrestrially.

Occurrence. Pyrrhotite is found in basic igneous rocks as scattered small grains or locally in greater concentrations, as at Sudbury, Ontario, where it is associated with pentlandite, chalcopyrite, and other sulphides. It also occurs in pegmatites, in contact-metamorphic rocks, and in hydrothermal veins. Pyrrhotite is found many places, e.g. at Dalnegorsk in Russia, Herja in Romania, Trepca in Serbia, and Santa Eulalia in Mexico.

Use. Pyrrhotite is of only local importance as an Fe ore and is primarily mined because of the associated Ni-rich sulphides, principally pentlandite.

Diagnostic features. Crystal habit, colour, hardness, and magnetic properties.

Figure 146. Pyrrhotite from Santa Eulalia, Chihuahua, Mexico. Field of view: 28 × 42 mm.

Figure 147. Pyrrhotite from Dalnegorsk, Russia. Field of view: 32 × 44 mm.

Nickeline, NiAs

Figure 148. Nickeline from Elk Lake, Timiskaming District, Ontario, Canada. Subject: 75 × 89 mm.

Crystallography. Hexagonal $6/m2/m2/m$; crystals uncommon, most often with a tabular habit; usually occurs massive.

Physical properties. No distinct cleavage, conchoidal or uneven fracture. Hardness 5½; density 7.8. Colour is light copper-red, tarnishing to grey or black; streak brownish-black; metallic lustre; opaque.

Chemical properties, etc. Ni can be replaced in small amounts by Fe or Co, and As by Sb. The structure is a hexagonal closest packing of spheres of As atoms, in which Ni is located in the positions octahedrally coordinated with As.

Names and varieties. The name *niccolite* was formerly used for nickeline. The mineral was first named *kupfernickel*, with reference to its colour, and that name in turn gave the name to the element Ni.

107

Figure 149. Millerite from Sterling Mine, Antwerp, New York, USA. Field of view: 36 × 54 mm.

Occurrence. Nickeline is found in basic igneous rocks and related ore bodies, commonly in association with pyrrhotite, chalcopyrite, and other Ni-containing sulphides and arsenides. It is also found in hydrothermal vein deposits, typically in association with minerals containing Co and Ag, as at Cobalt, Ontario, Canada.

Use. Nickeline is an inferior nickel ore.

Diagnostic features. Colour and hardness; is easily altered to the green mineral annabergite.

Breithauptite, NiSb

Hexagonal; structurally related to nickeline; occurs mostly massive. Properties closely resemble those of nickeline, but the colour has a more violet tinge and the streak is reddish-brown. Breithauptite is found in hydrothermal veins together with Co–Ni–Ag ore minerals; it is known, e.g. from Coyote Peak in California, USA, the Hemlo gold deposit in Ontario, Canada, and St. Andreasberg, Harz Mountains, Germany.

Millerite, NiS

Trigonal; crystals mostly slender to needle-shaped parallel to the *c* axis, commonly in radiating crystal groups or fibrous masses; rarely massive. Cleavage perfect on $\{10\bar{1}1\}$ and $\{01\bar{1}2\}$; brittle, needle-shaped crystals partly elastic. Hardness 3–3½; density 5.5. Light brass-coloured, streak black with a greenish tinge; metallic lustre; opaque. Millerite is a low-temperature hydrothermal mineral usually found crystallized in cavities and fissures in limestones or in calcite veins. It is also found as an alteration product replacing other Ni-containing minerals. Fine crystal groups have been found at several localities, e.g. Stirling Mine, Antwerp in New York, USA, and from waste tips of old coal mines in South Wales.

Pentlandite, $(Ni,Fe)_9S_8$

Crystallography. Cubic $4/m\bar{3}2/m$; crystals very rare; usually granular, massive; often intimately associated with pyrrhotite.

Physical properties. In larger grains some cleavage or parting on {111}, conchoidal fracture; brittle. Hardness 3½–4; density about 4.8. Light bronze-yellow, streak bronze-brown; metallic lustre, opaque.

Chemical properties, etc. The Ni/Fe ratio is normally close to 1. Co can be present in small amounts.

Occurrence. Pentlandite occurs in basic igneous rocks, commonly in association with other ore minerals containing Ni and Fe. The most significant occurrence of pentlandite is found at Sudbury, Ontario, Canada. It is also known, e.g. from Outokumpu and other places in Finland.

Figure 150. Pentlandite from Evje, Norway. Subject: 81 × 112 mm.

Use. Pentlandite is the principal Ni ore. Ni is used in stainless steel and various alloys, nickel-plating, catalysts, crucibles, and rechargeable batteries.

Diagnostic features. Closely resembles pyrrhotite but is non-magnetic.

Covellite, CuS

Crystallography. Hexagonal $6/m2/m2/m$; crystals uncommon, mostly as tabular crystals with hexagonal striation on {0001}; occurs usually massive or as coatings on other Cu minerals.

Physical properties. Cleavage perfect on {0001}, cleavage folia partly flexible. Hardness 1½–2; density 4.7. Colour blue to black, usually with a play of brass-coloured or deep red colours; streak black and shining; metallic lustre; opaque, but translucent in thin blades.

Chemical properties, etc. Fe can to a limited extent replace Cu.

Occurrence. Covellite is a secondary mineral formed by the alteration of other sulphides containing Cu. It occurs mainly in association with chalcocite, bornite, and chalcopyrite in the enriched zone of Cu deposits (see chalcocite). Fine crystals are found, e.g. at Butte, Montana, USA.

Use. Covellite is a minor Cu ore.

Diagnostic features. Colour and cleavage.

Figure 151. Covellite from Alghero, Sardinia, Italy. Subject: 51 × 65 mm.

Cinnabar, HgS

Crystallography. Trigonal 32; crystals common, commonly with rhombohedral habit or tabular parallel to {0001}; penetration twins on {0001} common; occurs mostly granular, massive, earthy, as incrustations or finely dispersed in rocks.

Physical properties. Cleavage perfect on {10$\bar{1}$0}. Hardness 2½; density 8.1. Colour is vermilion-red, brownish when impure; streak red; adamantine lustre, dull when impure or earthy; transparent to translucent.

Chemical properties, etc. Fe oxides, clays, or organic compounds commonly contaminate cinnabar; impure masses of cinnabar are sometimes called *hepatic cinnabar*.

Occurrence. Cinnabar is formed at relatively low temperatures and occurs in veins and impregnation zones in sediments, usually in relation to recent volcanic activity; it is typically found together with pyrite, antimonite, quartz, chalcedony, calcite, and other carbonates. It is known, e.g. from Almadén in Spain, Idria in Slovenia, New Almaden and New Idria in California, USA, and as exceptionally well-developed crystals from the Hunan Province, China.

Use. Cinnabar is the principal ore of Hg, which is used, e.g. in instruments such as thermometers and barometers, electrical ap-

Figure 153. Cinnabar, a penetration twin on {0001}, from China. Field of view: 22 × 38 mm.

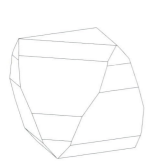

Figure 152. Cinnabar: {10$\bar{1}$0}, {0001}, {10$\bar{1}$1}, {02$\bar{2}$1}, and {20$\bar{2}$5}.

paratus, batteries, dental fillings, paints, and preservation media in agriculture.

Diagnostic features. Colour, streak, density, crystal habit, and cleavage.

Realgar, AsS

Crystallography. Monoclinic 2/*m*; crystals uncommon, usually in short prismatic crystals with striation parallel to the *c* axis; occurs mostly coarse to fine granular, massive or as incrustations.

Physical properties. Cleavage on {010} good, indistinct on {$\bar{1}$01}, {100}, and {120}; sectile. Hardness 1½–2; density 3.6. Colour is red to

111

orange-yellow, changing to a yellow mixture of arsenolite and orpiment when exposed to light; streak orange-yellow; resinous lustre; transparent when fresh.

Occurrence. Realgar is found in low-temperature hydrothermal veins in association with orpiment, stibnite, and other As–Sb minerals, commonly as a weathering product replacing other As minerals. It is also found as a volcanic sublimation product and in deposits at hot springs, as in Norris Geyser Basin, Yellowstone Park in Wyoming, USA.

Use. Both realgar and the related mineral orpiment have been used as pigments, but this practice has stopped because of their poisonous nature.

Diagnostic features. Colour, lustre, and hardness.

Orpiment, As_2S_3

Crystallography. Monoclinic $2/m$; crystals uncommon, mostly as short prismatic, pseudo-orthorhombic crystals; usually in foliated or columnar masses.

Physical properties. Cleavage perfect on {010}, cleavage lamellae flexible but inelastic; sectile. Hardness 1½–2; density 3.5. Lemon-coloured to brownish-yellow; streak pale yellow; resinous lustre, pearly on cleavage surfaces; translucent.

Occurrence. Orpiment occurs almost like realgar, commonly associated with realgar and as an alteration product of realgar. It is found, e.g. at Getschell mine and White Caps mine in Nevada, USA.

Use. Orpiment and realgar have been used as

Figure 154. Realgar from Allchar, Macedonia. Subject: 70 × 78 mm.

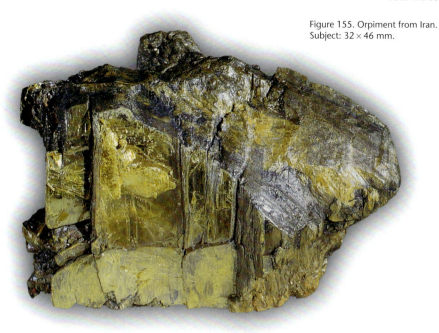

Figure 155. Orpiment from Iran. Subject: 32 × 46 mm.

pigments, but this practice has stopped because of their poisonous nature.

Diagnostic features. The foliated habit of aggregates, colour, lustre, and cleavage.

Stibnite, Sb_2S_3

Crystallography. Orthorhombic $2/m2/m2/m$; crystals common, mostly as long prismatic or needle-shaped crystals, sometimes bent or twisted, commonly arranged in radiated aggregates; prism faces usually striated or grooved parallel to the c axis; also occurs massive, coarse to fine granular.

Physical properties. Cleavage perfect on {010}, cleavage lamellae inelastic, often striated parallel to the a axis. Hardness 2; density 4.6. Lead-coloured to black, often weakly tarnished; streak lead-coloured to black; metallic lustre; opaque; dulls on exposure to light.

Chemical properties, etc. The composition of stibnite is normally close to the ideal formula.

Names and varieties. *Antimonite* is an obsolete name seen in old collections. *Antimonit* or *Antimonglanz* are German names for stibnite that also occur in a number of other European languages. *Bismuthinite*, Bi_2S_3, is a rare mineral related to stibnite and with similar physical properties.

Occurrence. Stibnite occurs in low-temperature hydrothermal veins, in replacement deposits, and in deposits at hot springs. It is usually associated with realgar, orpiment, galena, pyrite, baryte, quartz, and calcite. In surface conditions it is readily oxidized and decomposed to white or yellowish stibiconite. The largest deposits of stibnite are found in the Hunan Province in China. Well-developed crystal groups are known, e.g. from Baia Sprie in Romania, from Ichinokawa on Shikoku Island in Japan, and

113

Figure 156. Stibnite from Baia Sprie, Romania. Subject: 67 × 90 mm.

from San José and several other localities in Bolivia.

Use. Stibnite is the principal ore of Sb; substantial quantities of Sb are, however, derived as a by-product from Pb ores. Sb has been used as type metal, in various alloys, bearings and pigments.

Diagnostic features. Crystal habit, cleavage, and aggregate form. It has the same colour as galena but can be distinguished from it by having only one perfect direction of cleavage and a lower density.

Pyrite, FeS_2

Crystallography. Cubic $2/m\bar{3}$; crystals common, often as cube {100}, pentagonal dodecahedron {210}, octahedron {111}, or combinations of these; faces of {100} and {210} commonly striated parallel to the edge between faces of the two crystal forms as a result of oscillatory growth of these forms; twinning with [110] as twin axis is seen; also occurs massive, granular, and in radiating, globular, or stalactitic forms.

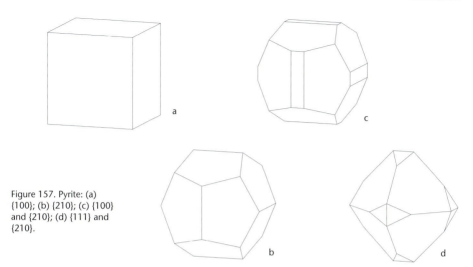

Figure 157. Pyrite: (a) {100}; (b) {210}; (c) {100} and {210}; (d) {111} and {210}.

Physical properties. No cleavage, conchoidal fracture; brittle. Hardness 6–6½; density 5.0. Brass-coloured, streak black with a greenish tinge; metallic lustre, opaque.

Chemical properties, etc. Pyrite and marcasite are polymorphs. Pyrite is normally close to its ideal formula; some Fe is occasionally replaced by Ni or Co.

The crystal structure of pyrite can be derived from that of halite, NaCl, by replacing Na by Fe and Cl by an S_2 group. The two S atoms are very close to each other and the bond between them is parallel to one of the four threefold axes. Fe is octahedrally coordinated with six S atoms.

Occurrence. Pyrite is a widespread mineral and the most common sulphide. It is found in almost all geological environments, as an accessory mineral in most igneous rocks, as a major mineral in many igneous ore bodies, in pegmatites, in contact-metamorphic rocks, and in hydrothermal veins. It is also widespread in many metamorphic rocks and sediments, e.g. as concretions in the English and Danish chalk deposits. Large concentrations of pyrite are known, e.g. from Rio Tinto in Spain. Very fine crystals are known from Quiruvilca and other mines in Peru. When exposed to oxidation, pyrite decomposes to Fe sulphates and Fe oxides. As a result, sulphuric acid is formed. In nature the acid affects the surrounding minerals; in museum and private collections the acid may destroy labels and boxes.

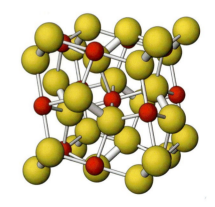

Figure 158. The crystal structure of pyrite can be derived from that of halite, NaCl, by replacing Na with Fe (red) and Cl with a pair of S atoms (yellow). The two S atoms are very close and the bond in between them is parallel to one of the four threefold axes of symmetry. Fe is octahedrally coordinated with six S atoms.

Figure 159. Pyrite from Navajun, Logroño, Spain. Subject: 48 × 44 mm.

Figure 160. Pyrite from Peru. Field of view: 48 × 44 mm.

Figure 161. Pyrite from Huanzala, Huanuco, Peru. Field of view: 66 x 99 mm.

Figure 162. Pyrite from Carbondale,
Illinois, USA. Diameter: 85 mm.

Figure 163. Pyrite from Frederiksholm Teglværk, South Harbour, Copenhagen, Denmark. Subject: 94 × 134 mm.

Use. Pyrite is of only minor importance as an Fe ore; it is mostly mined for the production of sulphuric acid and Fe sulphates, and occasionally because of its content of Cu and Au.

Diagnostic features. Crystal habit, colour, and hardness. Pyrite has a noticeable greater hardness than chalcopyrite and gold, and is also brittle in contrast to gold; pyrite usually resembles marcasite but generally has a more intense colour.

Marcasite, FeS_2

Crystallography. Orthorhombic $2/m2/m2/m$; crystals common, typically tabular parallel to {010} or with prismatic habit, twinning on {110} common; larger crystals commonly consist of several subparallel individuals or twins forming spear-shaped or cockscomb-like terminations; commonly in radiating forms; also stalactitic or botryoidal.

Physical properties. Cleavage distinct on {101}, conchoidal fracture, brittle. Hardness 6–6½; density 4.9. Light brass-coloured, almost white on fresh fracture faces; streak greyish-black; metallic lustre; opaque.

Chemical properties, etc. Marcasite, which is polymorphous with pyrite, normally deviates only slightly from the ideal formula. The crystal structure is similar to that of pyrite and the closest coordination around Fe is the same, but the S_2 pairs lie in planes perpendicular to the c axis.

Occurrence. Marcasite forms under low-temperature conditions and is mainly found in sediments and hydrothermal veins, in the latter case often in association with Pb and Zn ores, as in Joplin, Missouri, USA. It is less

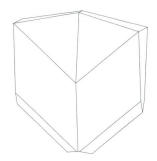

Figure 164. Marcasite: twinned on {110}; both individuals have {011}, {110}, and {014}.

Sperrylite, PtAs₂

Cubic; belongs to the pyrite group; crystals mostly {100}, {111}, or combinations thereof. Cleavage indistinct on {100}, conchoidal fracture; brittle. Hardness 6½; density 10.6. Colour is tin-white with a black streak; strong metallic lustre; opaque. Sperrylite is the commonest Pt mineral and is found primarily in ore deposits associated with pyrrhotite and pentlandite. It is known, e.g. from Sudbury in Canada, Norilsk in Russia, and Bushveld in the Republic of South Africa.

Cobaltite, CoAsS

Crystallography. Orthorhombic *mm*2; crystals pseudo-cubic with cube-like or pyritohedron-like forms, thus resembling pyrite; also granular.

stable than pyrite and readily decomposes to Fe sulphates and sulphuric acid.
Diagnostic features. Crystal habit and hardness; resembles pyrite but has a cleavage and a lighter colour.

Figure 165. Marcasite from Sokolov, Czech Republic. Subject: 60 × 106 mm.

119

Figure 166. Cobaltite from Håkansboda, Sweden. Subject: 40 × 38 mm.

Physical properties. Cleavage cube-like, more or less perfect; conchoidal or uneven fracture; brittle. Hardness 5½; density 6.3. Colour is silver-white with a reddish tinge, streak greyish-black; metallic lustre; opaque.

Chemical properties, etc. Co is frequently replaced in part by Fe, and to a lesser degree by Ni. The crystal structure is a pyrite structure, in which the S–S pair in pyrite is replaced by an S–As pair with a corresponding reduction in symmetry from cubic to orthorhombic.

Names and varieties. *Gersdorffite*, NiAsS, and *ullmannite*, NiSbS, are related minerals; they are relatively rare. *Linnaeite*, Co_3S_4, is a Co-rich sulphide known e.g. from Bastnäs in Sweden.

Occurrence. Cobaltite occurs in high-temperature hydrothermal veins in association with Ni and Co-rich ore minerals. It is also associated with some contact-metamorphic rocks. Cobaltite is known from Cobalt in Ontario, Canada, from Congo, the largest producer, and from, e.g. Håkansboda in Sweden, where large well-developed crystals have been found.

Use. Cobaltite is one of the most important Co ores. Co is primarily used in various steel alloys.

Diagnostic features. Crystal habit, colour, and cleavage.

Arsenopyrite, FeAsS

Crystallography. Monoclinic $2/m$, pseudo-orthorhombic; crystals common, often prismatic; twinning frequent, e.g. on {001}, contact and penetration twins e.g. on {101}, cru-

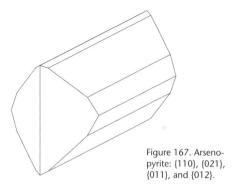

Figure 167. Arseno-pyrite: {110}, {021}, {011}, and {012}.

Chemical properties, etc. The As/S ratio can vary to a limited extent. The crystal structure of arsenopyrite can be derived from that of marcasite by replacing half the S with As.

Names and varieties. *Löllingite*, $FeAs_2$, is a closely related mineral that may be associated with arsenopyrite.

Occurrence. Arsenopyrite is the most common As mineral and the principal As ore. It occurs in medium- to high-temperature hydrothermal veins, is associated with gold in quartz veins and with, e.g. cassiterite in Sn-bearing veins; also with W minerals such as scheelite in contact-metamorphic rocks. It is widespread and is known, e.g. from Freiberg in Germany, Trepca in Serbia, Tavistock in Devon, tin mines in Cornwall, England, and Panasqueira in Portugal. Fine crystal groups are found in Santa Eulalia and other mines in Mexico. **Use.** Arsenopyrite is the principal ore of As. (For uses see *arsenic*.)

Diagnostic features. Crystal habit, including twin forms, and colour.

ciform twins and star-shaped trillings, e.g. on {012}; also granular, compact.

Physical properties. Cleavage distinct on {101}, indistinct on {010}; uneven fracture. Hardness 5½–6; density 6.1. Colour is silver-white to steel-grey, streak almost black; metallic lustre; opaque.

Figure 168. Arsenopyrite from Trepca, Serbia. Subject: 78 × 130 mm.

Molybdenite, MoS$_2$

Crystallography. Hexagonal $6/m2/m2/m$; crystals tabular on {0001}; occurs mostly foliated, scaly, or massive.

Physical properties. Cleavage perfect on {0001}, cleavage lamellae flexible but inelastic; feels greasy. Hardness 1–1½; density 4.7. Colour is lead-coloured with a bluish tinge, streak black with a greenish tinge; metallic lustre; opaque.

Chemical properties, etc. The chemical composition is normally close to the ideal formula. The crystal structure is layered and consists of composite layers in which a layer of Mo is sandwiched between two layers of S. The bonds within the composite layer are much stronger than those between these layers, which results in a perfect cleavage parallel with the layers.

Occurrence. Molybdenite is widespread but normally occurs only in small amounts. It is found as an accessory mineral in some granites and pegmatites, in pneumatolytic veins

Figure 170. Molybdenite from Kingsgate, New South Wales, Australia. Field of view: 40 × 50 mm.

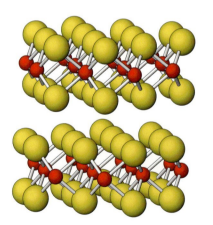

Figure 169. The crystal structure of molybdenite consists of composite layers in which a layer of Mo (red) is sandwiched between two layers of S (yellow). The bonds within the composite layer are much stronger than those between the layers, which results in a perfect cleavage parallel to the layers.

with, e.g. scheelite and wolframite, and in contact-metasomatic deposits. One of the most important deposits of molybdenite is at Climax in Colorado, USA. Fine crystals are found, e.g.at the Crown Point mine in Washington, USA, and at Kingsgate, New South Wales, Australia.

Use. Molybdenite is the principal ore of Mo, which is used for various purposes, e.g. in steel alloys, electrical components, and as a lubricant.

Diagnostic features. Cleavage and hardness; resembles graphite but is distinguished from it by a bluish tinge in its colour and a greenish tinge in its streak.

Sylvanite, AgAuTe₄

Monoclinic; crystals with prismatic or tabular habit, often complex; a different type of twinning results in threaded or dendritic forms resembling written characters (in German: *schrifterz*). Cleavage perfect on {010}; uneven fracture; brittle. Hardness 1½–2; density 8.2. Colour is silver-white, sometimes with a weakly golden tinge; streak grey; metallic lustre; opaque. Sylvanite occurs in low-temperature hydrothermal veins associated with other minerals rich in Ag and Au. It is known, e.g. from Sacaramb (formerly Nagyág), Romania, and other gold deposits in the old Transylvania, from which the mineral got its name. *Calaverite*, AuTe₂, and *krennerite*, (Au,Ag)Te₂, are closely related minerals and their occurrence is of the same type.

Skutterudite, (Co,Ni)As₃

Crystallography. Cubic $2/m\overline{3}$; crystals uncommon, mostly as {100}, {111}, or combinations of these, rarely with {110} or {hk0}; usually granular, massive.

Physical properties. Cleavage usually not distinct; fracture uneven or conchoidal; brittle. Hardness 5½–6; density 6.5. Colour is tin-white or steel-grey, at times coloured by tarnishing; streak black; metallic lustre; opaque.

Chemical properties, etc. Fe commonly replaces Co or Ni in part. Ni predominates over Co in nickel-skutterudite.

Names and varieties. Safflorite, CoAs₂, and rammelsbergite, NiAs₂, are closely related minerals commonly associated with skutterudite.

Occurrence. Skutterudite occurs in ore veins

Figure 171. Sylvanite from Baia-de-Arieș, Romania. Field of view: 56 × 84 mm.

Figure 172. Skutterudite from Schneeberg, Saxony, Germany. Subject: 76 × 74 mm.

together with cobaltite, nickeline, and other Co and Ni ore minerals. It is known from, e.g. the type locality Skutterud at Modum in Norway, Annaberg and Schneeberg in Saxony, Germany, and from Cobalt in Ontario, Canada.

Use. Skutterudite is mined as an ore of Co and Ni.

Diagnostic features. Crystal habit and colour are the most useful characteristics, but skutterudite is usually difficult to identify.

Sulphosalts

The sulphosalts are characterized by the fact that the semi-metals As, Sb, and Bi enter the metal positions in the crystal structure. In the sulphides and arsen-

ides, etc. described above these semi-metals play the same role as S and enter the same positions as S.

The sulphosalts are commonly minerals containing Ag, Cu, or Pb that occur in hydrothermal veins, usually in small amounts and generally in association with more common sulphides.

Pyrargyrite, Ag_3SbS_3

Crystallography. Trigonal $3m$; crystals usually prismatic or with a scalenohedron-like habit, can be very complex; commonly twinned on $\{10\bar{1}4\}$ or $\{10\bar{1}1\}$; also granular, massive.

Physical properties. Cleavage distinct on $\{10\bar{1}1\}$, conchoidal or uneven fracture; brittle. Hardness 2½; density 5.8. Colour is deep red, darkening with exposure to light; streak red; adamantine lustre; translucent.

Names and varieties. *Dark ruby silver* is an old name for pyrargyrite. *Stephanite*, Ag_5SbS_4, and *polybasite*, $(Ag,Cu)_{16}Sb_2S_{11}$, are other Ag- and Sb-containing sulphosalts.

Occurrence. Pyrargyrite is a late-stage mineral in low-temperature hydrothermal veins and is typically found with other Ag minerals, galena, and calcite. Among well-known deposits are those at St. Andreasberg, Harz Mountains, and Freiberg in Germany, Pribram in the Czech Republic, and Hiendelaencina in Spain.

Use. Pyrargyrite is an important Ag ore.

Diagnostic features. Crystal habit, colour, lustre, and mode of occurrence.

Figure 173. Pyrargyrite from Freiberg, Saxony, Germany.
Subject: 30×43 mm.

Proustite, Ag_3AsS_3

Crystallography. Trigonal $3m$; crystals as for pyrargyrite, although not usually as complex; twinning on $\{10\overline{1}4\}$ or $\{10\overline{1}1\}$ seen; also massive.

125

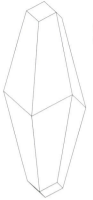

Figure 174. Proustite: {10$\bar{1}$1} and {32$\bar{5}$1}.

Figure 176. Tetrahedrite from Huacracocha, Junin, Peru. Field of view: 50 × 75 mm.

Physical properties. Cleavage distinct on {10$\bar{1}$1}, conchoidal or uneven fracture; brittle. Hardness 2–2½; density 5.8. Colour is scarlet-vermilion, somewhat lighter than pyrargyrite, darkening with exposure to light; streak vermilion; adamantine lustre; translucent.

Names and varieties. *Light ruby silver* is an old name for proustite.

Occurrence. Like pyrargyrite, proustite is a late-stage mineral in low-temperature hydrothermal veins and is typically found with other Ag minerals, galena, and calcite. It is generally associated with pyrargyrite but is less common.

Use. Proustite is locally an important Ag ore.

Diagnostic features. Crystal habit, colour, lustre, and mode of occurrence.

Tetrahedrite, $(Cu,Fe)_{12}Sb_4S_{13}$ – tennantite, $(Cu,Fe)_{12}As_4S_{13}$

Crystallography. Cubic $\bar{4}3m$; crystals usually tetrahedral; contact or penetration twins on {111} common; crystals sometimes overgrown by or in intergrowths with chalcopyrite; also massive, granular.

Physical properties. No cleavage, uneven fracture. Hardness 3–4, tennantite hardest; density about 4.8. Colour is grey to black; streak brown to black; metallic lustre; opaque.

Chemical properties, etc. There is a complete solid-solution series between tetrahe-

Figure 175. Tetrahedrite: {111}, {110}, {211}, and {21$\bar{1}$}.

drite and tennantite. Fe is always present to some extent and is therefore included in the formula, but Zn, Ag, Pb, and Hg can also be present; *freibergite* is an Ag-rich member of this group.

Occurrence. Tetrahedrite is probably the most common and economically most important sulphosalt. Both tetrahedrite and tennantite occur in hydrothermal veins, generally associated with other minerals containing Cu, Pb, Zn, and Ag. They are also found in contact-metamorphosed deposits. Good tetrahedrite crystals are found, e.g. in Park City District in Utah, USA, and in Noche Buena and other places in Zacatecas, Mexico.

Use. These minerals are important as Cu ores, and locally also as Ag ores.

Diagnostic features. Crystal habit and lack of cleavage.

Enargite, Cu_3AsS_4

Crystallography. Orthorhombic *mm*2, pseudo-hexagonal; crystals common, tabular parallel to {001} or prismatic with striations parallel to the *c* axis; twinning on {320} common, at times cyclic in star-shaped trillings; also granular, columnar.

Physical properties. Cleavage perfect on {110}, distinct on {100} and {010}; uneven fracture; brittle. Hardness 3; density 4.5. Colour is steel-grey to black, occasionally with a violet tinge; streak black; metallic lustre, tarnishing dull; opaque.

Chemical properties, etc. Sb and Fe can to a limited extent replace As and Cu respectively.

Occurrence. Enargite occurs in hydrothermal veins and replacement deposits, typically in association with pyrite, sphalerite, galena, and other Cu minerals such as chalcocite, chalcopyrite, bornite, and tetrahedrite. Locally it is of importance as a Cu ore. Well-developed crystals are known from, e.g. Butte, Montana, and various mines in the Red Mountain district, Colorado, USA.

Diagnostic features. Cleavage and striations on faces.

Figure 177. Enargite from Quiruvilca, La Libertad, Peru. Subject: 51 × 95 mm.

Bournonite, $CuPbSbS_3$

Orthorhombic; crystals usually short prismatic or tabular on {001}; twinning on {110}, often repeated; also massive, granular. Cleavage indistinct on {010}. Hardness 2½–3; density 5.8. Colour is steel-grey to black, streak black; metallic lustre; opaque. Bournonite is found in hydrothermal veins associated with other sulphosalts and sulphides. Particularly good crystals have been found in localities in Cornwall, England, such as Herodsfoot mine.

Boulangerite, $Pb_5Sb_4S_{11}$

Figure 178. Bournonite from Horhausen, Rheinland-Pfalz, Germany. Field of view: 20×19 mm.

Monoclinic; crystals uncommon, usually long prismatic to needle-shaped with striations parallel to the c axis; most often in fibrous masses. Cleavage good on {100}; brittle, thin fibres flexible. Hardness 2½–3; density about 6. Lead-coloured with a bluish tinge; streak brownish-grey; metallic lustre; opaque. Boulangerite is found in hydrothermal veins, usually with, e.g. galena and other sulphosalts containing Pb, and with stibnite, which it closely resembles. *Jamesonite*, $Pb_4FeSb_6S_{14}$, and *jordanite*, $Pb_{14}(As,Sb)_6S_{23}$, are other sulphosalts containing Pb.

Halides

Figure 179. Fluorite
with baryte (white)
and sphalerite (black)
from Elmwood Mine,
Tennessee, USA.
Subject: 135 × 170 mm.

The halide group of minerals includes simple compounds such as NaCl, in which a halogen element is bonded to an alkali metal. Fluorine (F), chlorine (Cl), and the much rarer halogens bromine (Br) and iodine (I) are present as large monovalent anions linked with pure ionic bonds to small mono- or divalent cations such as sodium (Na) or calcium (Ca). The bonding in such structures is relatively weak. Of special importance is the fact that the bonds are directed spherically towards the surroundings and not just in any direction. In consequence, the ions behave like almost perfect spheres and are packed with the highest possible symmetry. Many of these minerals are accordingly cubic, and they generally have a poor hardness.

Whereas the chlorides are mainly formed by evaporation from solutions, the fluorides are typically found in igneous rocks and their associated pegmatites and hydrothermal veins.

Halite, NaCl

Crystallography. Cubic $4/m\bar{3}2/m$; crystals common, mostly as simple cubes, sometimes with curved or stepped faces or as 'hopper crystals', i.e. skeletally developed; also in granular or compact masses.

Physical properties. Cleavage perfect on {100}. Hardness 2½; density 2.2. Colourless or white, also yellowish or reddish owing to impurities or bluish owing to radioactivity; vitreous lustre; transparent to translucent.

Chemical properties, etc. Halite is readily soluble in water and tastes salty. The crystal

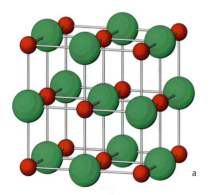
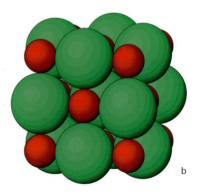

Figure 180. In the crystal structure of halite every Cl⁻ (green) is octahedrally coordinated with six Na⁺ (red), and every Na⁺ is octahedrally coordinated with six Cl⁻. The cube is here shown with Na⁺ at the corners, but it could equally well be shown with Cl⁻ at the corners. (a) A traditional drawing of the structure with ions, shown at less than their true size, in the lattice. It gives a good impression of the mutual placing of the ions but has the disadvantage that the lines between the ions may be misinterpreted as bonds. In a structure like this the bonds between an ion and its surroundings are directed spherically and not solely in one direction. (b) The structure without the lattice and with the ions in their correct proportions; it is magnified about 10 million times.

Figure 181. Halite from Inowrocław, Poland.
Field of view: 80 × 84 mm.

structure is a cubic lattice, in which every ion is octahedrally coordinated with six ions of the other kind.

Names and varieties. *Rock salt* is a term occasionally used for halite, usually in a petrographic context.

Occurrence. Halite is a widespread mineral. It is found as extended beds formed by the evaporation of sea water and usually alternates with beds of other salts such as gypsum, anhydrite, calcite, and, more rarely, sylvite and other K- or Mg-rich salts. Such series are of Palaeozoic to Recent age. The older sequences are commonly overlain by thick layers of sediments, the pressure of which has caused the rock salt to be deformed plas-

tically and to penetrate the overlying sediments as salt domes. Important deposits of this kind are found many places, e.g. in Germany, Austria, Poland, and in New York State, and the Gulf Coast region in the USA. Halite is also formed at salt springs and salt lakes in arid regions and as a sublimation product in volcanic areas.

Use. Halite is used for the production of hydrochloric acid and a great variety of other Na and Cl compounds. It is also used in the food industry, for domestic purposes, and for salting roads in winter.

Diagnostic features. Crystal form, cleavage, and salty taste (which, in contrast to sylvite, is not bitter).

Figure 182. Halite from Stassfurt, Germany. The blue colour is due to lattice defects, possibly caused by radioactivity. Subject: 70×74 mm.

Sylvite, KCl

Crystallography. Cubic $4/m\overline{3}2/m$; crystals common, usually {100} solely, but also in combination with {111}; mostly granular or compact.

Physical properties. Cleavage perfect on {100}; less brittle than halite. Hardness 2½; density 2.0. Colourless, white, or with light colours owing to impurities; vitreous lustre; transparent to translucent.

Chemical properties, etc. Sylvite is more soluble than halite and has a bitter salty taste. It has a halite crystal structure, but the miscibility between halite and sylvite is lim-

Figure 183. Sylvite from Stassfurt, Germany. Field of view: 60×58 mm.

Figure 184. Villiaumite from Alluaiv Mountain, Lovozero massif, Kola Peninsula, Russia. Subject: 83 × 125 mm.

ited owing to the great difference in ionic radius between Na^+ and K^+.

Occurrence. Sylvite has the same mode of occurrence as halite but is less common. It is formed as a late-stage mineral in the evaporation process because of its higher solubility as compared with halite. Extensive deposits of sylvite and other potassium salts are found at Stassfurt, Germany.

Use. Sylvite is used in the chemical industry for the production of various K compounds, especially fertilizers.

Diagnostic features. Very similar to halite but has a more bitter taste.

Villiaumite, NaF

Villiaumite has a halite structure and many properties in common with halite. It differs from halite in having a characteristic carmine-red colour and in its mode of occurrence; it is a late-stage mineral formed in nepheline-syenite complexes, e.g. Khibina and Lovozero in the Kola Peninsula, Russia, Ilímaussaq, Greenland, and Mont Saint-Hilaire, Quebec, Canada.

Chlorargyrite, AgCl

Crystallography. Cubic $4/m\bar{3}2/m$; crystal structure as halite; crystals rare, mostly as wax-like or horny masses or incrustations.

Physical properties. No cleavage, sectile. Hardness about 2; density 5.6. Colourless, grey, or yellowish, darkening by exposure to light to brown, violet, or black; streak shining; resinous to adamantine lustre when fresh, otherwise dull, translucent.

Figure 185. Chlorargyrite from Mercedes part Nudo, Tres Puntas, Chile. Subject: 62 × 96 mm.

Names and varieties. *Horn silver* and *cerargyrite* are obsolete names for chlorargyrite. *Bromargyrite*, AgBr, and *iodargyrite*, AgI, are closely related minerals. They are found with chlorargyrite but are less common.

Occurrence. Chlorargyrite is a supergene mineral found in the upper zones of Ag deposits, especially in arid regions. It was formerly an important Ag ore, mainly in the South American Andes.

Diagnostic features. Wax-like or horny appearance and sectility.

Sal ammoniac, NH$_4$Cl

Cubic; crystals small and rounded, often with {211}. No distinct cleavage; hardness 1½; density 2.0. Colourless, yellowish, or brownish; vitreous lustre; readily soluble in water. Sal ammoniac occurs as a sublimation product in volcanic areas and is known, e.g. from Vesuvius and Etna in Italy, Hekla in Iceland, and Kilauea, Hawaii, USA.

Figure 186. Sal ammoniac from Hekla, Iceland. Subject: 103 × 128 mm.

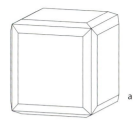
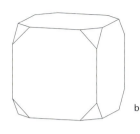
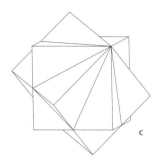

Figure 187. Fluorite: (a) {100} and {210}; (b) {100} and {111}; (c) a penetration twin with [111] as twin axis.

Fluorite, CaF_2

Crystallography. Cubic $4/m\bar{3}2/m$; crystals very common, most often {100}, more rarely {111}, {110}, {$hk0$}, or {hkl}, usually in combination with {100}; faces can vary in smoothness on same crystal, e.g. faces of {100} can be smooth, while faces of, e.g. {111} are rugged or stepped; penetration twins with [111] as twin axis common; also granular, rarely stalactitic.

Physical properties. Cleavage perfect on {111}. Hardness 4; density 3.2. Colour shows great variation and can be light green, blue-green, violet, yellow, or colourless, also pink, brown, or almost black; colours commonly zoned parallel to faces; colour can be caused by traces of hydrocarbons; vitreous lustre; transparent to translucent, perfectly transparent when pure.

Chemical properties, etc. The composition of fluorite is generally close to the ideal formula. In the crystal structure every Ca is co-ordinated with eight F at the corners of a cube, whereas F is tetrahedrally coordinated with four Ca.

Names and varieties. Fluorite has given its name to the property of fluorescence.

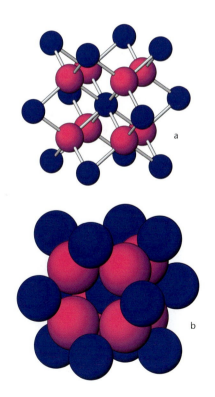

Figure 188. In the crystal structure of fluorite every Ca^{2+} (blue) is coordinated with eight F^- (red) at the corners of a cube, whereas F^- is tetrahedrally coordinated with four Ca^{2+}. The lines between the ions in (a) show the coordination of the ions and are not bonds. In a structure like this the bonding between an ion and its surroundings propagates spherically and not only in one particular direction. In (b) the structure with the ions in their correct proportions.

135

Figure 189. Fluorite from Geoda del Reguerin, La Collada, Asturias, Spain. Subject: 60 × 170 mm.

Occurrence. Fluorite is a widespread mineral found in many different associations. It is common in hydrothermal veins, where it is sometimes the principal mineral, and is typically associated with quartz, calcite, baryte, and various ore minerals. It is found in cavities and fissures in limestones, in pneumatolytically affected rocks, and in pegmatites. There are countless occurrences worldwide of well-developed fluorite crystals. Among well-

Figure 190. Fluorite from La Collada, Asturias, Spain. Subject: 60 × 64 mm.

Figure 191. Fluorite from Alston Moor, Cumbria, England. Field of view: 30 × 45 mm.

Figure 192. Fluorite from Saxony, Germany. Field of view: 41 × 66 mm.

Figure 193. Fluorite, stalactitic, from Ivigtut, Greenland. Field of view: 30 × 45 mm.

known localities in Europe are several in Cumbria, County Durham, and Derbyshire in England, Kongsberg in Norway, and the Asturia Province in Spain. Exceptional fluorite crystals are known from Weardale, County Durhjam, England, the Elmwood mine in Tennessee, USA, Huanzala in Peru, and Nagar in Pakistan, to name only a few.

Use. Fluorite is used as a flux in the smelting of aluminum and in steel production. It is also used in the production of many F compounds, e.g. freon. Particularly clear and flawless fluorite has been used for special optical purposes.

Diagnostic features. Crystal habit including twinning, cleavage, hardness, and colour.

Cryolite, Na_3AlF_6

Crystallography. Monoclinic $2/m$, pseudo-cubic; crystals rare, mostly cube-like with predominant {110} and {001}, commonly in groups of parallel oriented crystals grown on massive cryolite; twinning on several twin laws common; mostly massive or coarse granular.

Physical properties. No cleavage, parting on {110} and {001}, uneven fracture. Hardness 2½; density 3.0. Colourless or white, also brownish, reddish, or black; greasy vitreous lustre; transparent to translucent. When suspended in water cryolite looks like ice because of its low refractive indices.

Figure 194. Cryolite from Ivigtut, Greenland. Field of view: 110 × 165 mm.

Figure 195. Cryolite with siderite (brown) and sphalerite (black) from Ivigtut, Greenland. Subject: 114 × 140 mm.

Names and varieties. *Cryolithionite*, $Na_3Li_3Al_2F_{12}$, and *chiolite*, $Na_5Al_3F_{14}$, are examples of minerals that resemble cryolite and are associated with cryolite but are very rare.

Occurrence. The largest occurrence of cryolite was found at Ivigtut, Greenland, as a lens-shaped pegmatitic body in granite. It was associated with siderite, microcline, quartz, fluorite, and sulphides such as pyrite, chalcopyrite, sphalerite, galena, and pyrrhotite in addition to a number of very rare fluorides. This deposit was mined for more than 100 years but is now exhausted. Cryolite is also found, e.g. at Miask in the Urals, Russia, and at Pikes Peak in Colorado, USA.

Use. Natural cryolite was formerly used as a flux in the smelting of Al, a role now taken over by synthetic cryolite produced from fluorite.

Diagnostic features. Colour and lustre, parting, similarity to ice in water, and association with sulphides.

Pachnolite and thomsenolite

Polymorphs, both monoclinic with the formula $NaCaAlF_6 \cdot H_2O$. Both are colourless or superficially brownish, tinted by iron oxides;

Figure 196. Pachnolite from Ivigtut, Greenland.
The yellowish colour is caused by iron oxides.
Field of view: 30 × 45 mm.

Figure 197. Thomsenolite on stalactitic fluorite and
cryolite from Ivigtut, Greenland.
Field of view: 54 × 90 mm.

Figure 198. Atacamite from Copiapo, Atacama, Chile. Field of view: 35×52 mm.

vitreous lustre; hardness 3; density 3.0. Pachnolite is found as prismatic crystals with {110} and {111} and always twinned with the c axis as twin axis, whereas thomsenolite forms short to long prismatic crystals with {110} and {001}. Pachnolite has an indistinct cleavage on {001}, whereas thomsenolite has a perfect cleavage on {001} and a distinct cleavage on {110}. Both are found as crystals and as fine- to coarse-grained masses or coatings. They usually occur together and—especially in Ivigtut, Greenland—in close association with cryolite, normally as alteration products of cryolite.

Carnallite, $(K,NH_4)MgCl_3 \cdot 6H_2O$

Orthorhombic; crystals uncommon, mostly granular. No cleavage, conchoidal fracture; hardness about 2; density 1.6. Colourless,

Figure 199. Boléite from Boléo, Baja California, Mexico. Field of view: 24×36 mm.

milk-white, yellowish, or reddish owing to presence of hematite as impurity; dull greasy lustre; transparent to translucent; tastes bitter. Carnallite is found in marine evaporate salt deposits, usually in association with halite, sylvite, and kieserite. It is known, e.g. from Stassfurt, Germany.

Atacamite, $Cu_2Cl(OH)_3$

Orthorhombic; crystals mostly prismatic with striations parallel to the c axis; also columnar, granular, foliated, or fibrous. Cleavage perfect on {010}. Hardness 3–3½; density 3.8. Colour is bright green to deep green; streak apple-green; vitreous lustre; transparent to translucent. Atacamite is a supergene mineral occurring in the upper parts of Cu deposits in arid regions, e.g. in the Atacama Desert in Chile.

Boléite, $Ag_9Pb_{26}Cu_{24}Cl_{62}(OH)_{48}$

Cubic; crystals usually as {100} with subordinate {111}. Cleavage perfect on {100}, good on {110}; hardness 3–3½; density 5.1. Colour is Prussian blue, streak blue with a greenish tinge; vitreous lustre. Boléite is a supergene mineral known particularly from the Cu deposits at Boléo, Baja California in Mexico.

Figure 200. Rutile with hematite in quartz from Novo Horizonte, Bahia, Brazil. Subject: 59 × 70 mm.

Oxides and hydroxides

Oxides are compounds of oxygen (O) and one or more metals. They are normally built up as close packings of the large O atoms with the small metal atoms located in the interstices. The metals are usually coordinated with four or six O atoms. As a rule the chemical bond is a strong ionic bond, and oxides are generally characterized by great hardness and high density. They typically occur as accessory minerals in igneous and metamorphic rocks, and, owing to their great resistance to weathering and transport, also in sediments, where they can be concentrated in beds.

In hydroxides O is completely or partly replaced by OH-groups. Hydroxides are generally less hard and less dense than oxides, and occur typically in the upper weathering zone of ore deposits produced by the alteration of primary minerals.

Ice, H_2O, which is an oxide belonging to this group, is not included here. Silica, SiO_2, the most common of all oxides, is more closely related to the silicates and is therefore described under that heading.

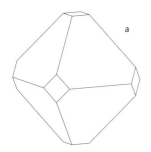

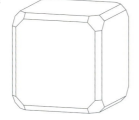

Figure 201. Cuprite:
(a) {100}, {111};
(b) {100}, {110},
and {111}.

Cuprite, Cu_2O

Crystallography. Cubic $4/m\overline{3}2/m$; crystals common, mostly {111}, {100}, {110}, and combinations thereof, at times in hair-like forms; also massive or earthy.

Physical properties. Cleavage distinct on {111}, uneven fracture, brittle. Hardness 3½–4; density 6.1. Colour is red in various shades to nearly black; streak reddish-brown; adamantine to submetallic lustre on fresh surfaces; almost opaque.

Chemical properties, etc. Cuprite is normally rather pure. In the crystal structure Cu is located midway between two O ions, and every O is tetrahedrally coordinated with four Cu.

Names and varieties. *Chalcotrichite* is a hair-like variety of cuprite, normally with a lighter red colour.

Occurrence. Cuprite is a supergene mineral occurring in the upper oxidized parts of copper deposits, usually associated with native

Figure 202. Cuprite from Mashamba, Shaba, Congo. Field of view: 35 × 43 mm.

copper, malachite, azurite, and limonite. Fine crystals are known from a number of localities, e.g. Bisbee in Arizona, Tsumeb and Emke in Namibia, and Mashamba in the Congo; in Europe from, e.g. Chessy in France and from several mines in Cornwall, England.

Use. Cuprite is a Cu ore of minor importance.

Diagnostic features. Crystal habit, colour, streak, and lustre. Cuprite is somewhat similar to hematite and cinnabar but is softer than hematite and harder than cinnabar.

Periclase, MgO

Cubic with a halite crystal structure; crystals rare, mostly tiny simple {111} or {100}; mostly as irregular or rounded grains. Cleavage perfect on {100}; hardness 5½–6; density 3.6. Colourless to yellowish-brown or green; vitreous lustre; transparent. Periclase is found in

Figure 203. Cuprite from Mashamba, Shaba, Congo. Field of view: 32 × 48 mm.

Figure 204. Periclase from Vesuvius, Italy. Subject: 54 × 75 mm.

Figure 205. Manganosite from Långban, Värmland, Sweden. Subject: 72 × 70 mm.

147

marbles formed by contact-metamorphorphism of dolomitic limestones and is known, e.g. from Monte Somma at Vesuvius, Italy. It is also known from Mn deposits at Långban in Sweden in association with the closely related mineral *manganosite*, MnO.

Zincite, ZnO

Hexagonal; crystals rare, mostly massive, granular or foliated. Cleavage perfect on $\{10\bar{1}0\}$, parting on $\{0001\}$. Hardness 4; density 5.7. Colour is red to orange-yellow; streak orange-yellow; resinous lustre; almost opaque. Zn is normally replaced to a moderate extent by Mn, which is considered to be the main cause of colour, since pure ZnO is white. Zincite is rare and was known in great-

Figure 206. Zincite from Franklin, New Jersey, USA. Subject: 86 × 123 mm.

Figure 207. Tenorite from Vesuvius, Italy. Subject: 61 × 68 mm.

er quantities only from deposits now exhausted at Franklin and Stirling Hill, New Jersey, USA, where it was found together with calcite, willemite, and franklinite.

Tenorite, CuO

Monoclinic; crystals rare and tiny; commonly as coatings of fine soot-like powder. Cleavage not seen; hardness 3½; density 6.4. Colour is greyish-black; metallic lustre; almost opaque. Tenorite is known as thin coatings on lavas at Vesuvius in Italy and as a weathering product in the upper oxidized zones of Cu deposits, usually in association with cuprite.

The spinel group

The oxides of the spinel group are of the type AB_2O_4, in which A is a divalent metal, e.g. Mg^{2+}, Fe^{2+}, Zn^{2+}, or Mn^{2+}, and B a trivalent metal, e.g. Al^{3+}, Fe^{3+}, or Cr^{3+}. Most spinels fall into three series determined by the B metal: a spinel series with Al^{3+}, a magnetite series with Fe^{3+}, and a chromite series with Cr^{3+} as the B metal. There is extensive solid solution within each series, whereas solid solution between the three series is limited.

Spinels are cubic and have a structure that is almost a cubic closest packing of

149

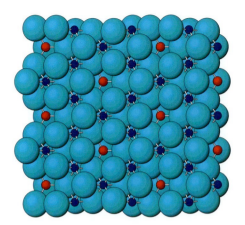

Figure 208. The crystal structure of spinel-group minerals can be considered as an almost perfect cubic closest packing of spheres of oxygen anions (light blue), with the small metal ions located in cavities between the large anions. The structure is here viewed along a threefold axis of symmetry. Three layers of anions are shown. The anions are drawn at less than their true size in order to reveal the sites of the cations. In the mineral spinel, Al (dark blue) is octahedrally coordinated with six O, and Mg (red) is tetrahedrally coordinated with four O. The coordination is indicated by bond symbols; the red Mg hides a bond symbol pointing backwards.

Spinel, $MgAl_2O_4$

Crystallography. Cubic $4/m\overline{3}2/m$; crystals common, often {111} solely or modified by {110} or {100}; twinning on {111}, the spinel law, common, twins commonly slightly tabular parallel to {111}; also massive, granular.

Physical properties. No cleavage, parting indistinct on {111}, conchoidal fracture. Hardness 7½–8; density 3.6, increasing with the Fe, Zn, or Mn content. Colour is red when pure, but also blue, green, brown, colourless, or black; vitreous lustre; transparent to translucent.

Chemical properties, etc. There exist almost complete solid-solution series between spinel and the end-members *hercynite*, $FeAl_2O_4$, *galaxite*, $MnAl_2O_4$, and *gahnite*, $ZnAl_2O_4$. Al can be replaced to a limited extent, primarily by Fe or Cr. Spinel has the spinel structure described above.

Names and varieties. *Pleonaste* is a Fe-rich black spinel found in peridotites and other basic rocks.

spheres of oxygen atoms. The metals are located in the interstices between these large anions, and their distribution is such that the *A* metals are tetrahedrally coordinated with four anions and the *B* metals are octahedrally coordinated with six anions. This description applies to normal spinels, but the inverse spinels have a different distribution of the metal ions, as described under magnetite.

As the packing of the anions is dense and the bonds to the di- and trivalent ions are strong, the spinels are generally of high density and great hardness.

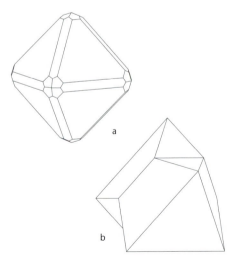

Figure 209. Spinel: (a) {110}, {111}, and {311}; (b) twinned on {111}, the spinel law.

Figure 210. Spinel from Gou, Yakutia, Russia. Field of view: 36 × 41 mm.

Occurrence. Spinel occurs as an accessory mineral in basic igneous rocks, in Al-rich metamorphic rocks, and in contact-metamorphic limestones. It is occasionally found in veins and pegmatites. As spinel is resistant to both chemical and physical weathering it is also found in sand and gravel deposits. Exceptionally large and well-crystallized spinels have been found at localities near Betroka, Madagascar.

Use. Transparent and fine-coloured spinels are prized gemstones; most stones come from gravel deposits in the Mogok region in Myanmar (Burma) and in Sri Lanka.

Diagnostic features. Crystal habit, hardness, lustre, and lack of cleavage.

Gahnite, $ZnAl_2O_4$

Gahnite, a member of the spinel group, is dark green in colour with a vitreous lustre. It is a rare mineral primarily found in metamorphic rocks, e.g. at Falun in Sweden, where it occurs in gneiss in association with sphalerite, galena, and other sulphides. It is known also from the Zn deposits at Franklin, New Jersey, USA.

Magnetite, Fe_3O_4

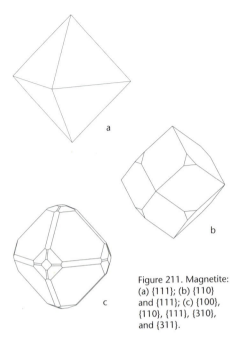

Crystallography. Cubic $4/m\overline{3}2/m$, belongs to the spinel group; crystals common, usually {111}, more rarely {110} or {100}; twinning according to the spinel law, {111}, common; also massive, fine to coarse granular.

Physical properties. No cleavage, distinct parting on {111} common, conchoidal fracture. Hardness about 6; density 5.2. Colour and streak black; metallic lustre, dull to bright; opaque; strongly magnetic.

Chemical properties, etc. Magnetite has the spinel structure described above. Magnetite belongs to the inverse spinels, in which the distribution of the di- and trivalent metals is different from that in normal spinels. In magnetite, the formula for which can be written as $Fe^{2+}Fe^{3+}_2O_4$, half the Fe^{3+} ions are in tetrahedral coordination, whereas the other half of the Fe^{3+} and all the Fe^{2+} ions are in octahedral coordination. Fe^{2+} can partly be

Figure 211. Magnetite: (a) {111}; (b) {110} and {111}; (c) {100}, {110}, {111}, {310}, and {311}.

Figure 212. Magnetite on tremolite from the Gardiner complex, Greenland. Field of view: 90 × 135 mm.

Figure 213. Magnetite from Binntal,
Switzerland. Field of view: 26 × 39 mm.

replaced by Mg^{2+}, Zn^{2+}, or Mn^{2+}, whereas Fe^{3+} can, to a lesser degree, be replaced by Al^{3+}, Cr^{3+}, Mn^{3+}, or Ti^{4+}.

Names and varieties. Closely related minerals are *magnesioferrite*, $MgFe_2O_4$, known from fumaroles at Vesuvius in Italy; *jacobsite*, $MnFe_2O_4$, known from Jakobsberg and Långban in Sweden; and *ulvöspinel*, Fe_2TiO_4, known from Ulvön in Sweden, which is commonly found as exsolution lamellae in magnetite. *Martite* is a pseudomorph of hematite after magnetite.

Occurrence. Magnetite is one of the most abundant oxide minerals. It is dispersed as an accessory mineral in most igneous rocks, occasionally concentrated in magmatic segregated masses. It is also found in many meta-morphic or sedimentary rocks, where it can form thicker layers, bands, or lens-shaped bodies; in metasomatically altered lime-stones, where it is associated with sulphides and Ca- and Fe-rich silicates; in high-temperature veins, in association with corundum in emery deposits, and concentrated in sands. One of the most important occurrences of magnetite is at Kiruna and Malmberget in Sweden, where it is associated with apatite. There are numerous localities with well-crystallized magnetite in Europe, e.g. Binntal in Switzerland and Zillertal in Austria. Strong natural magnets are found, e.g. in the Harz Mountains in Germany and on Elba, Italy. In the USA fine crystals are found, e.g. at the Tilly Foster mine in New York State.

Use. Magnetite is an important Fe ore.

Diagnostic features. Crystal habit, colour, streak, hardness, and magnetic properties.

Figure 214. Chromite from Texas, Lancaster County, Pennsylvania, USA. Subject: 78×98 mm.

Franklinite, $(Zn,Mn,Fe)(Fe,Mn)_2O_4$

Franklinite belongs to the spinel group and resembles the other spinels; crystals are rare, commonly rounded; usually massive. Colour is black or brownish-black; streak reddish-brown; metallic or submetallic lustre; opaque; weakly magnetic. As indicated in the formula, the chemical variation is considerable. Franklinite was the most important Zn ore of the deposits in Franklin and Stirling Hill, New Jersey, USA.; minerals typically associated include calcite, zincite, and willemite.

Chromite, $FeCr_2O_4$

Crystallography. Cubic $4/m\bar{3}2/m$, belongs to the spinel group; crystals rare, usually {111}; mostly fine granular, massive.

Physical properties. No cleavage, uneven fracture. Hardness 5½; density 4.6. Colour is black; streak brown; metallic lustre; opaque.

Chemical properties, etc. There are solid-solution series between chromite and the end-members *hercynite*, $FeAl_2O_4$ and *magnesiochromite*, $MgCr_2O_4$; an essential part of the Fe in chromite is commonly replaced by Mg, and a minor part of the Cr by Al or Fe^{3+}. Chromite has a spinel structure, described above.

Occurrence. Chromite occurs mostly as an accessory mineral in basic or ultrabasic igneous rocks such as peridotites and in serpentinites derived from these rocks. It occurs dispersed or in segregated bands or lenses, which can be valuable ores, as in the Bushveld complex in the Republic of South Africa. Chromite is also found in sands.

Use. Chromite is the only ore of Cr, which is used for chromium-plating, in stainless-steel alloys, refractories, and pigments, and in the tanning of leather.

Diagnostic features. Mode of occurrence, including association with olivine or antigorite. The streak distinguishes chromite from magnetite.

Maghemite, $Fe_{2.67}O_4$

Cubic; composition close to Fe_2O_3, but as it has a spinel-like structure the formula is written in a special way, as above. Maghemite is brown with a brown streak and is usually found as impure, fine-grained masses. It is as strongly magnetic as magnetite. Maghemite occurs as an alteration product of Fe-containing minerals such as siderite and magnetite, and is widespread in laterites in tropical regions.

Hausmannite, Mn_3O_4

Tetragonal with a distorted spinel structure; crystals uncommon, mostly in octahedron-like simple {101} bipyramids; repeated twinning on {112} seen; usually massive or granu-

Figure 215. Hausmannite from N'chwaning Mine, Kuruman, Republic of South Africa. Subject: 73×101 mm.

Figure 216. Chrysoberyl from Madagascar. Subject: 22 × 22 mm.

lar. Cleavage perfect on {001}, uneven fracture; hardness 5½–6; density 4.8. Colour is brown to black, streak chestnut-brown; submetallic lustre, nearly opaque. Hausmannite occurs in high-temperature hydrothermal veins, in contact-metamorphic deposits, and in Mn-rich sediments. It is known, e.g. from Långban and Jakobsberg in Sweden, from Ilfeld and Öhrenstock in Germany, and from the Kuruman manganese deposits in the Republic of South Africa.

Chrysoberyl, $BeAl_2O_4$

Crystallography. Orthorhombic $2/m2/m2/m$; crystals common, commonly tabular parallel to {001} with striation parallel to the *a* axis; twinning on {130} common, often repeated to form pseudo-hexagonal trillings.

Physical properties. Cleavage distinct on {110}, indistinct on {010}. Hardness 8½; density 3.7. Colour is green to yellowish-green or brownish, and can alternate between green and red (alexandrite); vitreous lustre; transparent.

Chemical properties, etc. Chrysoberyl has a crystal structure like that of olivine, in which Be replaces the tetrahedrally coordinated Si and Al the octahedrally coordinated Mg. O ions form an almost hexagonal closest packing of spheres, which explains the pseudo-hexagonal character of chrysoberyl.

Names and varieties. *Alexandrite* is a gemstone variety of chrysoberyl; it is green in daylight but red under tungsten light. *Cat's eye* is a variety in which needle-shaped inclusions give an effect resembling a cat's eye, which is best seen when cut *en cabochon*.

Occurrence. Chrysoberyl occurs in granites and granite pegmatites, in mica-schists, and in sands and gravels in association with other resistant minerals such as diamond and corundum. The largest gravel deposits with chrysoberyl are found in Sri Lanka and Brazil, whereas the most prized alexandrite gems come from a mica-schist near the Takowaja River in the Urals, Russia. Large crystals are found near Golden in Colorado, USA.

Use. Chrysoberyl is a prized gemstone.

Diagnostic features. Hardness, colour, crystal habit, and twinning.

Valentinite, Sb_2O_3

Orthorhombic; found as single prismatic crystals with many faces or crystals assembled in radiating or fan-shaped bundles; also columnar, granular, massive. Cleavage perfect on {110} and {010}; hardness 2–3; density 5.8. Colourless, white, grey, or yellowish to brownish; adamantine lustre; transparent to translucent. Valentinite occurs as a weathering product of Sb ores, especially stibnite. *Senarmontite*, Sb_2O_3, a polymorph of valentinite, occurs in the same way but is less widespread. It is cubic and is found as octahedra, but otherwise has properties resembling

Figure 217. Valentinite from Pribram, Bohemia, Czech Republic. Field of view: 30×37 mm.

Figure 218. Bixbyite from Thomas Range, Juab County, Utah, USA. Field of view: 21 × 23 mm.

those of valentinite. *Stibiconite*, $Sb_3O_6(OH)$, is a related mineral also found as a weathering product of other Sb minerals, especially stibnite; it is found as a white or yellowish-brown powder or as earthy incrustations. *Arsenolite*, As_2O_3, crystallizes as senarmontite, and is mainly found as white coatings on arsenopyrite and other As sulphides.

usually predominates over Fe, in which case the mineral should strictly be given a different name because at the type locality, Thomas Range, Utah, USA, Fe predominates over Mn. At the type locality bixbyite is found in cavities in rhyolite associated with topaz, garnet, and red beryl.

Bixbyite, $(Fe,Mn)_2O_3$

Cubic; crystals usually {100}, modified by {211}. Cleavage on {111}; hardness 6½; density 5.0. Colour and streak are black; metallic lustre; opaque. The Fe/Mn ratio varies; Mn

Corundum, Al_2O_3

Crystallography. Trigonal $\bar{3}2/m$; crystals common, often barrel-shaped with several steep hexagonal bipyramids and {0001}, perhaps also {11$\bar{2}$0}; also tabular parallel to {0001}; crystals commonly rough with un-

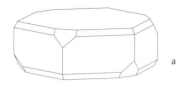

Figure 219. Corundum: (a) {0001}, {11$\bar{2}$0}, {10$\bar{1}$2}, and {11$\bar{2}$3}; (b) {0001}, {10$\bar{1}$2}, {11$\bar{2}$1}, {11$\bar{2}$3}, and {77$\bar{14}$3}.

even faces; prism and pyramid faces sometimes striated parallel to {0001}; striation also on {0001} parallel with prism faces; twinning common on {10$\bar{1}$1} and {0001}, usually lamellar; also granular.

Physical properties. No cleavage, parting on {0001} and {10$\bar{1}$1}. Hardness 9; density 4.0. Colour is usually grey, weakly blue, yellow, or red, but all colours can occur; vitreous lustre; transparent to translucent. Needle-shaped inclusions can give special optical effects, as in star sapphire, in which a sixfold star-shaped figure plays on a polished surface approximately perpendicular to the principal axis.

Chemical properties, etc. Corundum is normally pure Al_2O_3; the colour-giving ions, such as Cr (ruby) or Fe and Ti (sapphire), are present only in very small amounts, i.e. a few parts per million. The crystal structure can be described as an almost hexagonal close packing of spheres of O atoms in which two-thirds of the octahedrally coordinated interstices are occupied by Al and the remaining third is empty.

Names and varieties. *Ruby* is the red gemstone variety of corundum. *Sapphire* includes all other coloured gem varieties. Sapphire

with no specification of colour is blue; other varieties are called *yellow sapphire, white sapphire*, etc. *Emery* is a fine-grained mixture of corundum and magnetite or hematite known, e.g. from the Greek island of Naxos. **Occurrence.** Corundum occurs in Si-poor igneous rocks such as syenites and nepheline-syenites and associated pegmatites, in contact zones between peridotites and surrounding rocks, and in metamorphic rocks such as gneisses, mica-schists, and crystalline limestones. Because of its hardness and chemical resistance it is also widespread in sand and gravel deposits.

The finest rubies come from the Mogok region in Myanmar (formerly Burma), where they are found in metamorphosed limestones as well as in the overlying weathered zones. Particularly beautiful blue sapphires are known from deposits at high altitude at Padar in Kashmir, India, where marble and other metamorphic rocks are cut by pegmatites. Most of the precious corundums derive from gravel deposits in Sri Lanka, Cambodia,

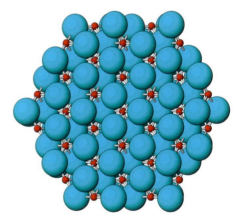

Figure 220. The crystal structure of corundum is an almost perfect hexagonal closest packing of oxygen anions (light blue) in which two-thirds of the octahedrally coordinated cavities are occupied by Al (red) while one-third remain empty. The structure is viewed along the threefold axis of symmetry; it shows three layers of anions and the two layers of cations between them. The anions are drawn at less than their true size in order to show the location of the cations.

Figure 221. Corundum var. ruby from Froland near Arendal, Norway. Subject: 73 × 80 mm.

and Thailand, but there are also important localities in Madagascar, Tanzania, and Queensland in Australia.

Use. Corundum is used as an abrasive, both when pure and when impure (as emery). A large proportion of the material used nowadays as abrasives is, however, produced synthetically from bauxite. Precious corundums are sought after; especially fine rubies are much appreciated and can achieve prices surpassed only by emerald and some diamonds. Many rubies are produced synthetically, and such stones can be difficult to distinguish from natural ones.

Diagnostic features. Hardness, parting, and crystal habit, especially the barrel-shaped habit. The surface of a crystal can be unexpectedly soft because of the presence of a thin layer of an alteration product of corundum.

Figure 222. Corundum var. ruby from India. Field of view: 11 × 15 mm.

Figure 223. Corundum var. sapphire from Sri Lanka. Subject: 33×111 mm.

Hematite, Fe_2O_3

Crystallography. Trigonal $\overline{3}2/m$; crystals uncommon, mostly tabular parallel to {0001} in variable thickness, sometimes in rosettes, {0001} faces with triangular striations; twinning on {10$\overline{1}$1} and {0001}, commonly lamellar; usually found in scaly, fibrous, or radiating masses, reniform, botryoidal, stalactitic, granular, concretionary, oolitic, or earthy.

Physical properties. No cleavage, parting on {0001} and {10$\overline{1}$1}; conchoidal fracture. Hardness about 6 in non-earthy forms; density 5.3. Colour is steel grey to black, at times with bluish iridescence, fine-grained varieties usually red or brown; streak reddish-brown; metallic lustre, earthy forms dull; opaque; non-magnetic.

Chemical properties, etc. Hematite is normally rather pure; only minor amounts of Ti, Mn, or H_2O can be present. The crystal structure of hematite is that of corundum with Fe^{3+} replacing Al.

Names and varieties. *Hematite* (formerly *haematite*) is derived from the Greek word for blood: a reference to the colour of the mineral when powdered. *Specular hematite* is hematite in tabular crystals with a distinct metallic lustre; *kidney ore*, reniform of botryoidal hematite, is an important variety. *Martite* is a pseudomorph of hematite after magnetite.

Figure 224. Corundum var. star sapphire from Australia. Subject: 29×29 mm.

161

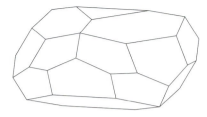

Figure 225. Hematite: {0001}, {01$\bar{1}$2}, {01$\bar{1}$8}, and {11$\bar{2}$3}.

Occurrence. Hematite is abundant in various geological associations and is the main mineral in most significant iron-ore deposits. It is found dispersed in relatively small amounts in many igneous rocks, in contact- metamorphic deposits, in high-temperature hydrothermal veins, and as a sublimate from volcanic activity. It is found dispersed in small amounts as well as in greater concentrations in regional metamorphic rocks, in which it has typically been formed by alteration of other Fe ores. It is also dispersed in many sediments and soils, where it is the prevalent colouring pigment. Hematite is especially widespread in banded iron formations, in which it constitutes extended layers alternating with quartz-rich layers. Such deposits are locally enriched in Fe by partial dissolution of the associated quartz bands or by metamorphism, e.g. by contact with a granitic intrusion. Many sedimentary occurrences of hematite are believed to have been formed by bacterial activity, which causes Fe hydroxides to be precipitated from hydrous solutions and subsequently dehydrated.

Huge deposits of hematite are found in several places, e.g. around Lake Superior, USA, and at Itabira in Brazil. Elba, Italy and St. Gotthard in Switzerland are among localities especially known for fine crystals of hematite. Congonhas do Campo in Minas Gerais is another of many localities for fine hematite in Brazil. Cumbria in England is one of the finest localities for 'kidney ore'.

Use. Hematite is the most important Fe ore. Fe is the chief constituent of steel, cast iron, and various alloys, and is thus one of the principal raw materials of modern civilization.

Diagnostic features. Streak; form of aggregate can be diagnostic; not magnetic, in contrast to magnetite.

Figure 226. Hematite from Montreal Mine, Montreal, Wisconsin, USA. Subject: 76 × 182 mm.

Figure 227. Hematite from St. Gotthard, Switzerland. Subject: 110 × 218 mm.

Ilmenite, $FeTiO_3$

Crystallography. Trigonal $\bar{3}$; crystals uncommon, typically tabular parallel to {0001} with one or two hexagonal prisms or rhombohedra; twinning on {10$\bar{1}$1} and {0001}; usually massive, granular; also in sands.

Physical properties. No cleavage, parting on {0001} and {10$\bar{1}$1}; conchoidal fracture. Hardness about 6; density 4.8. Colour is black or nearly black; streak black; metallic or submetallic lustre; opaque; non-magnetic or only weakly magnetic.

Chemical properties, etc. To a limited extent Fe can replace Ti, and Mg and Mn can replace Fe. The crystal structure of ilmenite can be derived from that of corundum by replacing Al_2 with FeTi. Fe and Ti are ordered in alternating layers, which results in a reduction of symmetry from $\bar{3}2/m$ to $\bar{3}$.

Occurrence. Ilmenite is a common accessory mineral in igneous rocks; it occurs segregated in greater masses in gabbros, diorites, and anorthosites, commonly in association with magnetite. It is an important mineral in sand deposits, where it is commonly associated with magnetite, zircon, rutile, and other minerals. It is also found in veins and pegmatites. Most of the ilmenite that is mined is from heavy sand deposits in Australia and several other places in the southern hemisphere. Crystals of ilmenite are known from, e.g. the Ilmen Mountains in Russia and Kragerø in Norway, where ilmenite is found in veins in diorite. Large crystals are found in the Lake Sanford area in New York State, USA.

Use. Ilmenite is the principal Ti ore. Ti is used as a construction material in aircraft and rockets, for bone prostheses, and for other applications in which high strength must be combined with low density; the oxide is used as a pigment in colours and glazes.

Diagnostic features. Streak distinguishes ilmenite from hematite and magnetic properties distinguish ilmenite from magnetite.

Perovskite, $CaTiO_3$

Crystallography. Orthorhombic $2/m2/m2/m$, pseudo-cubic; crystals usually cube-like, also octahedron-like, faces commonly striated as a result of repeated twinning; also granular.

163

Figure 228. Ilmenite from Arendal, Norway. Subject: 38 × 46 mm.

Physical properties. Cleavage not distinct, uneven fracture. Hardness 5½; density 4.0. Colour is black, rarely brown or yellow; grey or no streak; metallic or adamantine lustre; mostly opaque.

Chemical properties, etc. The chemical composition can vary to some extent: Na, Fe, or Ce can replace Ca, and Nb replace Ti. In *loparite*, Ce predominates over Ca.

Figure 229. Perovskite from Zlatoust, Urals, Russia. Field of view: 34 × 62 mm.

Figure 230. Perovskite from the Gardiner complex, Greenland. Field of view: 55 × 78 mm.

Occurrence. Perovskite is found as an accessory mineral in basic igneous rocks, in larger concentrations in some strongly differentiated igneous rocks such as carbonatites, and in limestones metamorphosed in contact with alkaline or basic intrusions. It is also found in chlorite- and talc-schists. It is believed that the greater part of the earth's mantle consists of minerals with crystal structures of the perovskite type. Fine perovskite crystals are known, e.g. from a chlorite-schist near Zlatoust in the Urals, Russia, and from the Gardiner complex in Greenland, where large crystals are found in a basic intrusion.

Diagnostic features. Mode of occurrence, lustre, and striation of crystal faces.

Braunite, $Mn^{2+}(Mn^{3+})_6SiO_{12}$

Tetragonal; crystals rare, usually pyramidal with {101} and {311}, twins on {112} seen; mostly granular, massive. Cleavage perfect on {112}, uneven fracture. Hardness 6–6½; density 4.8. Colour is brownish-black to steel grey; streak grey to black; metallic lustre; opaque; weakly magnetic. Braunite occurs especially in veins and lenses formed by metamorphism of Mn-rich mineral associations and subsequent weathering. It is typically associated with other Mn oxides, such as hausmannite and pyrolusite. Braunite is known, e.g. from the Spiller Mn mines in Texas, USA.

Figure 231. Pyrochlore from Taseq, Ilímaussaq complex, Greenland. Subject: 85 × 104 mm.

Pyrochlore, $(Na,Ca)_2Nb_2(O,OH,F)_7$

Crystallography. Cubic $4/m\overline{3}2/m$; crystals most commonly in octahedra; usually granular, massive.

Physical properties. Cleavage good on {111}, uneven fracture. Hardness 5–5½; density about 4.5. Colour is brown, with a yellowish or reddish tinge, to black; vitreous lustre, somewhat resinous; mostly opaque.

Chemical properties, etc. Pyrochlore is the most common of a group of minerals with the formula $A_2B_2(O,OH,F)_7$, where A can, e.g. be Na, Ca, Y, Ce, Sr, Ba, U, or Pb, and B Nb, Ta, or Ti, as in *microlite*, $(Ca,Na)_2Ta_2O_6$ $(O,OH,F)_7$ and *betafite*, $(Ca,Na,U)_2(Ti,Nb)_2O_6$ $(O,OH,F)_7$. There are extended solid-solution series between these minerals, and most of the minerals contain some U or Th, which has made them metamict.

Occurrence. Pyrochlore occurs in carbonatites and in alkaline pegmatites, typically in association with zircon, apatite, and minerals with Ce or other rare earth-elements. Pyrochlore is known, e.g. from Fen and the region around Langesundsfjorden in Norway, from Alnö in Sweden, St. Peter's Dome in Colorado, USA, and Oka in Quebec, Canada.

Use. Pyrochlore is an important ore of Nb. The largest deposits are found in association with carbonatites.

Diagnostic features. Crystal habit and mode of occurrence.

Rutile, TiO_2

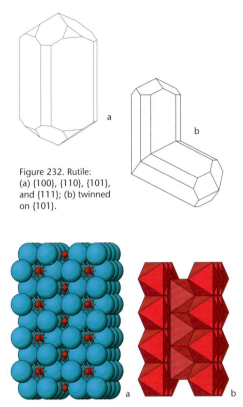

Crystallography. Tetragonal $4/m2/m2/m$; crystals common, usually prismatic, long prismatic, or needle-shaped, frequently with both {100} and {110}, and terminated with {101} or {111}; prism faces often striated parallel to the c axis; twinning common on {101}, frequently repeated in cyclic forms, e.g. to eight-membered rings; also granular.

Physical properties. Cleavage distinct on {110}, indistinct on {100}, uneven fracture. Hardness 6–6½; density 4.2. Colour is reddish-brown, golden brown, red, or black; streak light brown; adamantine or submetallic lustre; transparent in thin fragments, otherwise translucent.

Figure 232. Rutile: (a) {100}, {110}, {101}, and {111}; (b) twinned on {101}.

Chemical properties, etc. To a small extent Ti can be replaced by Fe, Nb, or Ta. This has to happen in combination because the valencies of these elements differ from that of Ti. In rutile Ti is in octahedral coordination, i.e. it is surrounded by six O atoms. These $[TiO_6]$ octahedra share edges and thus form chains parallel to the c axis; the chains are linked together by the octahedra sharing corners. In this way every O atom is bonded to three Ti atoms, which form a triangle that is almost equilateral. The existence of chains along the c axis is reflected in the pronounced prismatic or needle-shaped habit of rutile crystals. Rutile is polymorphous with anatase and brookite.

Occurrence. Rutile is a common accessory mineral in metamorphic rocks such as gneisses, mica-schists, and eclogites, and in igneous rocks such as granite and syenite. It is also found in pegmatites and quartz veins, and in addition is an essential constituent of sands. Rutile is commonly seen as needle-shaped inclusions in quartz crystals, sometimes as six-fold radiating crystals growing from a core of hematite. Well-crystallized rutile is known from many localities; some of the best known are in the Ouro Preto district and Novo Horizonte, Brazil, Graves Mountain in Georgia, Stony Point in North Carolina, and White Mountains in California, USA. Fine crystals are also found in Binntal, Switzerland, and other places in the Alps.

Figure 233. In the crystal structure of rutile Ti (red) is in octahedral coordination, i.e. it is surrounded by six O atoms (light blue). These $[TiO_6]$ octahedra share edges and form chains parallel to the c axis; the chains are linked by sharing corners of the octahedra, so that every O atom is bonded to three Ti. In (a) the ions are drawn at less than their true size in order to show their positions. In (b) the structure is shown in simplified form with $[TiO_6]$ octahedra only, so that the chains are more recognizable. The presence of chains along the c axis is reflected in the pronounced prismatic or needle-shaped habit of rutile.

Use. Rutile is primarily used as a coating on welding rods and as a source for the metal Ti.

Diagnostic features. Crystal habit, including twinning, lustre and colour.

Figure 234. Rutile from Novo Horizonte, Bahia, Brazil. Subject: 40 × 44 mm.

Figure 235. Rutile from Azerbaijan. Field of view: 38 × 57 mm.

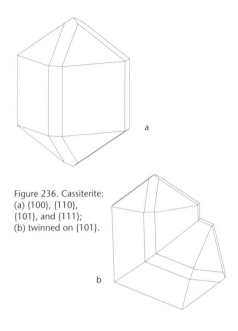

Figure 236. Cassiterite:
(a) {100}, {110},
{101}, and {111};
(b) twinned on {101}.

especially found in pneumatolytic veins in association with granitic complexes, in which it is typically associated with wolframite, scheelite, molybdenite, arsenopyrite, tourmaline, or topaz. It is also found concentrated in sand deposits, as, e.g. in Malaysia, Thailand, and Indonesia; these deposits are the major tin ores. Cornwall in England was previously famous for its tin mines, but today Bolivia is the only country with an important production from vein deposits. Among the famous localities for fine cassiterite crystals are, in addition to Cornwall and several places in Bolivia, Horni Slavkov (formerly Schlaggenwald), and other occurrences in the Saxonian–Bohemian Erzgebirge.

Use. Cassiterite is the principal Sn ore. Sn is used, e.g. for tinplating, bronze and other alloys, and solders.

Diagnostic features. Density, hardness, lustre, and crystal habit.

Figure 237. Cassiterite from an unknown locality. Field of view: 27×37 mm.

Cassiterite, SnO_2

Crystallography. Tetragonal $4/m2/m2/m$; crystals common, habit varies, usually short prismatic with {100}, {110}, and the bipyramids {101} and {111}; {111} sometimes prevailing, {101} commonly heavily striated; twinning very common on {101}, often repeated; occurs usually granular, massive, fibrous, radiated, or botryoidal.

Physical properties. Cleavage indistinct on {100}, uneven fracture. Hardness 6–7; density 7.0. Colour is reddish-brown to brownish-black, rarely yellowish; streak white or light yellow; adamantine to metallic lustre; mostly translucent.

Chemical properties, etc. Sn can to a limited extent be replaced by Fe, more rarely by Nb or Ta. Cassiterite has a rutile crystal structure in which Sn replaces Ti.

Occurrence. Cassiterite, one of the few Sn minerals, is widespread in small amounts but rarely occurs in greater concentrations. It is

Plattnerite, α-PbO_2

Tetragonal with a rutile structure; crystals rare, mostly needle-shaped; usually in dense, fibrous or botryoidal masses. No cleavage; hardness 5½; density 9.6. Colour is black to brownish-black; brown streak; metallic to adamantine lustre, tarnish dull; opaque. Plattnerite is found in the upper oxidized zones of Pb ore deposits and is known, e.g. from Leadhills in Scotland and in large masses in the Gilmore and other districts in Idaho, USA.

Pyrolusite, MnO_2

Crystallography. Tetragonal $4/m2/m2/m$; crystals uncommon; mostly in radiating or fibrous masses that are solid or earthy and often soil the fingers, concretionary, as coatings, dendritic; as pseudomorph from manganite.

Physical properties. Cleavage perfect on {110}, splintery or uneven fracture. Hardness highly variable, from 6–6½ on coarse crystalline faces to 1–2 for earthy forms; density 4.4–5.1. Colour is steel-grey or iron-black, commonly with a bluish tinge; streak black; metallic lustre; opaque.

Chemical properties, etc. Massive forms usually contain a small amount of water. Pyrolusite has a rutile structure.

Names and varieties. *Polianite* is an obsolete name for coarsely crystalline pyrolusite. *Wad* is a term used in the field for impure mixtures of Mn oxides, often hydrated, almost equivalent to the term *limonite* for hydrated Fe oxides.

Occurrence. Pyrolusite is one of the most abundant Mn minerals. It forms under highly oxidizing conditions and is found as an alteration product of other Mn-containing minerals everywhere in weathering zones, including various ore deposits, and it is usually associated with other oxides and hydroxides

Figure 238. Pyrolusite from Rossbach, Rheinland-Pfalz, Germany. Field of view: 53 × 97 mm.

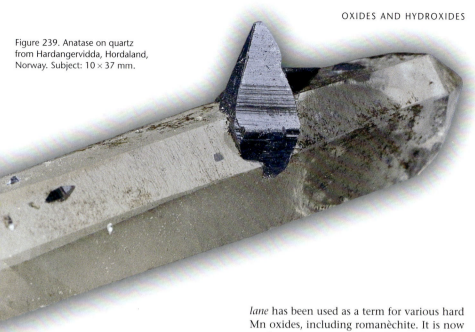

Figure 239. Anatase on quartz from Hardangervidda, Hordaland, Norway. Subject: 10 × 37 mm.

containing Mn or Fe. Extensive deposits of nodular pyrolusite have been found on ocean floors, and also in bogs, lakes, and shallow waters.

Use. Pyrolusite is the principal Mn ore. Manganese is used, e.g. in steel and other alloys, electric batteries, to decolourize glass, and as an oxidizing medium in the production of various chemicals.

Diagnostic features. Streak and, especially in relation to other Mn oxides and hydroxides, hardness (may soil the fingers).

Romanèchite, $(Ba,H_2O)_2Mn_5O_{10}$

Orthorhombic; occurs mostly as botryoidal incrustations, in stalactitic or earthy masses. Hardness 5–6; density 4.7. Colour and streak black to brownish-black; metallic lustre; opaque. Romanèchite forms under surface conditions and has much the same mode of occurrence as pyrolusite. *Cryptomelane*, KMn_8O_{16}, is one of a series of closely related minerals not distinguishable from romanèchite without special investigations. *Psilome-lane* has been used as a term for various hard Mn oxides, including romanèchite. It is now used as a general term for hard, massive Mn oxides (cf. *wad*). χ-psilomelane has been used specifically as a synonym for romanèchite.

Anatase, TiO_2

Crystallography. Tetragonal $4/m2/m2/m$; crystals usually acute bipyramidal predominated by {101}, rarely tabular parallel to {100}.

Physical properties. Cleavage perfect on {001} and {101}, uneven fracture. Hardness 5½–6; density 3.9. Colour is bluish-black, also yellow, brown, reddish-brown, rarely colourless, green, blue, or grey; streak whitish; adamantine lustre, metallic when dark; transparent to translucent.

Chemical properties, etc. Anatase is polymorphic with rutile and brookite.

Occurrence. Anatase is more rare than rutile. It is typically found in Alpine veins in gneisses and schists, often associated with quartz, adularia, and brookite; it is also known from pegmatites, and as dispersed grains in sediments. Among well-known localities are Beaver Creek in Colorado with blue crystals, Binntal in Switzerland (yellow crystals), and

171

Brookite, TiO$_2$

Orthorhombic, polymorphic with rutile and anatase; crystals tabular parallel to {010} or prismatic with predominant {120}. Cleavage indistinct on {120}; hardness 5½–6; density 4.1. Colour is yellowish-brown, reddish-brown to black; metallic or adamantine lustre; transparent to translucent. Brookite is found in Alpine veins, usually in association with anatase, rutile, and albite; fine crystals are known from Fron Oleu near Tremadoc, Wales, from several localities in the Alps, e.g. Bourg d'Oisans, France, and from Magnet Cove, Arkansas, USA.

Columbite, (Fe,Mn)(Nb,Ta)$_2$O$_6$

Crystallography. Orthorhombic $2/m2/m2/m$; crystals common, mostly prismatic, commonly tabular parallel to {010}; twinning on {201} common.

Figure 241. Brookite from Magnet Cove, Arkansas, USA. Field of view: 26×29 mm.

Figure 240. Anatase on quartz from Hardangervidda, Hordaland, Norway. Subject: 38×111 mm.

Hardangervidda in Norway (bluish-black crystals on quartz).
Diagnostic features. Crystal habit and mode of occurrence.

172

Figure 242. Columbite from Tammela, Finland. Field of view: 37 × 42 mm.

Physical properties. Cleavage distinct on {010}, indistinct on {100}, uneven fracture. Hardness 6–6½; density 5.2–6.8, rising with increasing Ta content. Colour is black to brownish-black, sometimes tarnished; streak

brown to black; metallic lustre; translucent to opaque.

Chemical properties, etc. There is a solid-solution series between *ferrocolumbite*, $(Fe,Mn)(Nb,Ta)_2O_6$, and *manganocolumbite*, $(Mn,Fe)(Nb,Ta)_2O_6$; most columbites have Fe > Mn. Both Fe and Mn are divalent. These minerals also form solid-solution series with *manganotantalite*, $(Mn,Fe)(Ta,Nb)_2O_6$, and *ferrotantalite*, $Fe(Ta,Nb)_2O_6$; tantalites usually have Mn > Fe. Pure ferrotantalite is not found in nature, but a close relative, *ferrotapiolite*, is known, e.g. from the Kulmala pegmatite at Tammela, Finland.

Occurrence. Columbite occurs in granitic pegmatites, more rarely in alkali granitic pegmatites. It is found, e.g. at Haddam in Connecticut and in several mines in South Dakota, USA. Large manganotantalite crystals are known form Amelia, Virginia, USA

Diagnostic features. Crystal habit, hardness, tarnishing colours.

Aeschynite, $(Ce,Ca,Fe)(Ti,Nb)_2(O,OH)_6$

Orthorhombic; occur as rough prismatic crystals or massive. Hardness 5–6; density 4.2–5.3. Colour is dark brown; streak brown to black; metallic lustre; usually metamict

Figure 243. Aeschynite from Hidra (formerly Hitterø), Norway. Subject: 72 × 127 mm.

Figure 244. Fergusonite from Evje, Norway. Subject: 29 × 45 mm.

owing to presence of U or Th. The chemical composition can vary considerably; e.g. the Ti/Nb ratio varies and Ce can be replaced by other rare-earth elements or Y. *Blomstrandine* and *priorite* are obsolete names for aeschynite. Aeschynite and closely related minerals occur in granite or nepheline-syenite pegmatites, in carbonatitic rocks, and in Alpine veins. Aeschynite is known, e.g. from Hidra (formerly Hitterø), Norway. *Polycrase*, $Y(Ti,-Nb)_2(O,OH)_6$, and *euxenite*, $Y(Nb,Ta,Ti)_2O_6$, are other related minerals; they are normally also metamict owing to minor amounts of U or Th replacing Y. Their occurrence is similar to that of aeschynite.

Fergusonite, $YNbO_4$

Tetragonal; crystals prismatic or acute pyramidal; also granular. Hardness 5–6½; density 4.2–5.7. Colour is usually black or brownish-black; submetallic lustre; opaque; normally metamict owing to presence of U or Th. The chemical composition varies: Y is replaced by Ce or other rare-earth elements, and Nb is replaced by Ti or Ta. Fergusonite occurs in granitic pegmatites and is known from, e.g. Baringer Hill in Texas, Amelia in Virginia, USA, and Madawaska in Ontario, Canada. *Samarskite*, $(Y,Fe,U)(Nb,Ta)O_4$, is a related mineral also found in granitic pegmatites.

Figure 245. Uraninite from Råde, Norway. Subject: 12 × 11 mm.

Uraninite, UO_2

Crystallography. Cubic $4/m\bar{3}2/m$; crystals usually simple combinations of {100} and {111}; mostly massive as *pitchblende* or in dense reniform, botryoidal, or banded masses.

Physical properties. No cleavage, uneven or conchoidal fracture. Hardness 5–6; density about 11, decreasing to about 6.5 with increasing geological age, by oxidation of U^{4+} to U^{6+} and replacement of U by Th and other elements. Colour is black to brownish-black; streak brownish-black; submetallic or pitchy lustre, mostly dull; opaque. Strongly radioactive.

Chemical properties, etc. In nature uraninite is generally oxidized and has a composition closely corresponding to U_3O_8. Th and rare-earth elements such as Ce can enter in variable amounts; in the case of Th to a considerable degree. Pb and He are always present as a result of the radioactive decay of U and Th. Uraninite has a fluorite structure, in which U replaces Ca and O replaces F.

Names and varieties. *Thorianite*, ThO_2, has a structure like that of uraninite; it is mostly found in granitic pegmatites.

Occurrence. Uraninite occurs in granitic and syenitic pegmatites, where it is typically associated with monazite, zircon, and oxides rich in Nb, Ta, Ti, and rare-earth elements. Deposits of this type are found, e.g. at Bancroft in Ontario, Canada, and several places in the south of Norway. It also occurs in high-tem-

Figure 246. Uraninite from Section 25 Mine, McKinley County, New Mexico, USA. Subject: 64 × 116 mm.

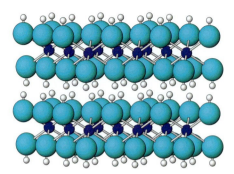

Figure 247. Brucite has a layered structure in which every layer consists of a layer of Mg (dark blue) sandwiched between two layers of OH groups (light blue and white). Mg is in octahedral coordination, surrounded by six OH groups. The OH groups form a hexagonal closest packing of spheres, and each of them is bonded to three Mg on one side and to three OH-groups in the neighbouring layer on the other side. The OH groups are polarized by having H atoms (white) located opposite to the Mg layer. The bonds between the composite layers are mostly hydrogen bonds.

perature hydrothermal veins with cassiterite and arsenopyrite, as in Cornwall, England. It is found as pitchblende in hydrothermal veins formed at moderate temperatures, together with Co–Ni–Ag ores, as, e.g. at Jachymov in the Czech Republic and at Great Bear Lake in Saskatchewan, Canada; an occurrence of a similar type but with a Cu ore is found at Shinkolobwe in the Congo. Finally, uraninite is found as small grains in quartz conglomerates, as for instance in the Au-bearing Witwatersrand conglomerate in the Republic of South Africa. Uraninite in pegmatites normally appears as crystals, whereas in hydrothermal veins it is mostly found as pitchblende. Fine crystals are known from, e.g. Topsham in Maine and Spruce Pine in North Carolina, USA, and large crystals from Wilberforce, Ontario, Canada.

Use. Uraninite is the principal ore of U and Ra. U is primarily used in nuclear reactors, whereas Ra is used in medicine for radiotherapy.

Diagnostic features. Density and pitch-like appearance.

Brucite, $Mg(OH)_2$

Crystallography. Trigonal $\bar{3}2/m$; crystals rare, usually tabular parallel to {0001} and encircled by a rhombohedron, single crystals or in subparallel aggregates; usually in foliated, massive or fibrous masses.

Physical properties. Cleavage perfect on {0001}, cleavage folia flexible but inelastic, sectile. Hardness 2½; density 2.4. Colour is white to weakly green, also grey, brown, or blue; wax-like vitreous lustre, pearly lustre on cleavage surfaces; transparent.

Chemical properties, etc. Divalent Fe and Mn can replace Mg. Brucite has a layered crystal structure in which each layer consists of two layers of OH groups embracing a layer of Mg. Mg is in octahedral coordination surrounded by six OH. The OH groups constitute a hexagonal closest packing of spheres, and each of them is bonded to three Mg on one side and to three OH in the next layer. The OH groups are polarized by H opposite to the Mg layer. The bonds between the composite layers are largely hydrogen bonds. The layered structure is reflected in the perfect cleavage.

Occurrence. Brucite occurs in low-temperature hydrothermal veins in serpentinites,

Figure 248. Brucite in calcite from Tilly Foster mine, Brewster, New York, USA. Subject: 41 × 43 mm.

Figure 249. Bauxite from France. Subject: 76 × 95 mm.

chlorite-schists, and dolomitic rocks, usually in association with calcite, aragonite, magnesite, and talc. It is found as an alteration product of periclase in crystalline limestones. Large brucite crystals are found in Wood's Mine, Texas, Pennsylvania, USA, and particularly fine crystals are known from a number of occurrences in the Urals, Russia.

Use. Brucite is used in refractories and as a source of Mg.

Diagnostic features. Foliated appearance, inelastic cleavage folia, and, to some extent, colour and lustre. Resembles talc but is harder and feels less greasy. Fibrous brucite is not as silky as chrysotile.

Gibbsite, $Al(OH)_3$

Crystallography. Monoclinic $2/m$; crystals rare, usually pseudo-hexagonal, tabular parallel to {001}; usually occurs in radiated aggregates, stalactitic, as incrustations, or earthy.

Physical properties. Cleavage perfect on {001}; Hardness 2½–3½; density 2.4. Colour is white to grey, also light greenish or reddish; vitreous lustre, pearly on cleavage surfaces; transparent.

Chemical properties, etc. Gibbsite has a structure similar to that of brucite, but with Al in the octahedral positions instead of Mg; only two-thirds of the positions are occupied in gibbsite, whereas all are occupied in brucite.

Names and varieties. *Hydrargillite* is an obsolete name for gibbsite (and also for aluminite).

Occurrence. Gibbsite is one of the three important constituents of bauxite; the other two are böhmite and diaspore. *Bauxite* is formed by supergene processes in subtropical or tropical regions. It is derived from the disintegration of Al-rich rocks that are leached of Si and other main elements except Al. The bauxite can remain as a residual deposit or be transported and secondarily deposited. *Laterite* is a similar residual deposit; it is found in tropical regions and is a reddish-brown earthy mixture of hydroxides containing Al and Fe. Bauxite is known from Les Baux in Provence and several other places in southern France, and from many other countries, e.g. Surinam, Guyana, and Jamaica. Gibbsite is also found in low-temperature hydrothermal veins and in cavities in alkaline igneous rocks. Richmond in Massachesetts and White Mountains in California, USA, are among the prominent localities for gibbsite.

Use. Bauxite is the principal ore of Al, which is used in aircraft, warships, cars, food containers, and for many other purposes where a light metal is required.

Diagnostic features. Gibbsite is less hard than diaspore; fine granular gibbsite is not easily distinguished from the other minerals present in bauxite.

Figure 250. Manganite: {010}, {101}, {210}, {111}, {810}, {212}, and {818}.

Figure 251. Manganite from Ilfeld, Harz Mountains, Germany. Subject: 55×93 mm.

Figure 252. Diaspore from Chester, Massachusetts, USA. Subject: 55 × 94 mm.

Manganite, MnO(OH)

Crystallography. Monoclinic $2/m$, pseudo-orthorhombic; crystals uncommon, mostly medium to long prismatic parallel to the c axis and with striation parallel to this axis; twinning on {011} common; usually columnar or fibrous.

Physical properties. Cleavage perfect on {010}, good on {110} and {001}. Hardness 4; density 4.3. Colour is steel-grey to black; streak dark brown to black; submetallic lustre; opaque.

Occurrence. Manganite occurs in low-temperature hydrothermal veins, typically associated with baryte, calcite, siderite, and hausmannite. It is also found in weathering zones, commonly with pyrolusite and other oxides containing Mn. Particularly fine crystals are known from Ilfeld, Harz Mountains, and Ilmenau, Thüringen, in Germany, Powell's Fort near Woodstock, Virginia, USA, and from St. Just, Cornwall in England.

Diagnostic features. Crystal habit, colour and, especially in relation to pyrolusite and other Mn oxides, hardness and streak.

Diaspore, AlO(OH)

Crystallography. Orthorhombic $2/m2/m2/m$; crystals rare, most often long prismatic or needle-shaped and tabular parallel to {010}; usually as dispersed grains, massive, scaly, or stalactitic.

Physical properties. Cleavage perfect on {010}, distinct on {110}. Hardness 6½–7; density about 3.4. Colour is white to grey, also colourless, greenish, brownish or reddish; vitreous lustre, pearly lustre on cleavage surfaces; transparent to translucent.

Chemical properties, etc. The crystal structure of diaspore is of the goethite type with Al replacing Fe^{3+}.

Names and varieties. *Böhmite*, AlO(OH), is polymorphous with diaspore, and like diaspore is an essential constituent of bauxite. It cannot be distinguished from the other Al oxides without elaborate investigations.

Occurrence. Diaspore is one of the three important constituents of bauxite; the others are gibbsite and böhmite. (For bauxite, see gibbsite). It is also found in crystalline limestones, in emery associated with corundum,

Figure 253. Goethite crystals from Lake George, Park County, Colorado, USA. Subject: 69 x 125 mm.

and as a hydrothermal mineral in alkaline pegmatites.
Use. As constituents of bauxite, diaspore, and böhmite are important Al ores.
Diagnostic features. Cleavage and hardness.

Goethite, FeO(OH)

Crystallography. Orthorhombic $2/m2/m2/m$; crystals uncommon, mostly prismatic along the c axis and striated parallel to it; also tabular parallel to {010} or needle-shaped; occurs mostly in reniform, botryoidal, or stalactitic masses, commonly with internal fibrous or concentric development; also pisolitic, oolitic, or earthy in more-or-less porous masses.
Physical properties. Cleavage perfect on {010}, good on {100}. Hardness 5–5½; density

Figure 254. Goethite var. bog iron ore, from Helga Sjön, Kronoberg, Sweden. Diameter: 17–24 mm.

Figure 255. Pisolitic goethite from Solingen, Germany. Subject: 71 × 103 mm.

Figure 256. Goethite var. bog iron ore from Mecklenburg, Germany. Subject: 70 × 91 mm.

3.3–4.4. Colour is dark brown, yellowish-brown, or reddish-brown; streak yellowish-brown; adamantine to submetallic lustre, most often dull, fibrous masses silky; translucent.

Chemical properties, etc. Mn often replaces Fe to a small extent. The crystal structure of goethite is a hexagonal close packing of O atoms and OH-groups, in which Fe is in octahedral coordination.

Names and varieties. *Lepidocrocite*, FeO(OH), is a relatively rare polymorph of goethite.

Occurrence. Goethite is one of the most abundant minerals and is the main constituent of what in the field is usually called 'limonite'. It forms as an alteration product (analogous to rust) of other Fe-containing minerals under oxidizing conditions, and constitutes the main mass of the cap or 'gossan' covering many ore deposits. It is also an essential constituent of laterite and similar residual deposits. It is precipitated directly by inorganic or biogenic processes in lakes and bogs, as *bog iron ore*. Deposits of goethite can be valuable as Fe ores; e.g. the minette ore in Alsace-Lorraine, France, and some laterite deposits in Cuba. Fine crystals are known, e.g. from Pikes Peak in Colorado, USA, and St. Just in Cornwall, England.

Use. Goethite is an important Fe ore.

Diagnostic features. Goethite is distinguished from hematite by its streak.

Figure 257. Malachite (green) and azurite (blue) from Bisbee, Arizona, USA. Field of view: 40 × 60 mm.

Carbonates, nitrates, and borates

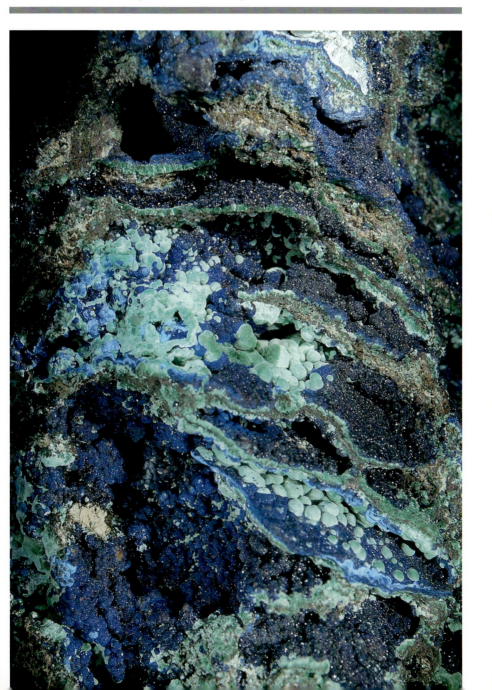

Carbonates are compounds of $[CO_3]$ groups and one or a few cations. A $[CO_3]$ group is a complex anion, which consists of a plane group with a carbon atom in the centre and three oxygen atoms placed tightly round it at the corners of an equilateral triangle. The $[CO_3]$ groups are isolated in the structure and do not form chains, rings, or layers such as are found in silicates. Similar conditions are found in nitrates, whereas borates resemble silicates.

Carbonates, primarily calcite and dolomite, are important minerals in sediments such as chalk and limestone, and in metamorphic rocks like marble. They are also widespread in hydrothermal veins. Nitrates are readily soluble and are found only under very dry conditions; borates are mostly formed by evaporation from salt lakes.

The calcite group

The crystal structure of calcite, $CaCO_3$, can be derived from that of halite, NaCl, by replacing Na with Ca and Cl with a $[CO_3]$ group. The replacement of a spherical unit such as Cl by a plane structural unit such as a $[CO_3]$ group reduces the symmetry from cubic to trigonal, be-

cause only one of the four threefold axes is preserved. This change can be described morphologically by saying that a cube has been changed to an obtuse rhombohedron. In analogy with this, the cleavage on a cube in halite is changed to cleavage on an obtuse rhombohedron in calcite, i.e. it is still in three directions but in calcite they are oblique

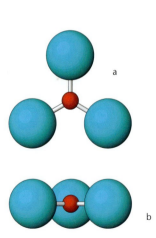

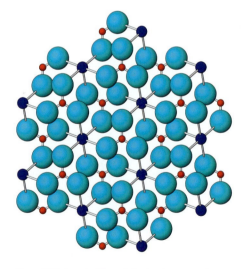

Figure 258. A $[CO_3]$ group seen (a) along the c axis and (b) perpendicular to it. C (red) is placed in the middle of a equilateral triangle with O (light blue) at each corner. The O atoms are drawn at less than their true size in relation to C in order to reveal the structure.

Figure 259. A part of the calcite structure seen along the c axis. C (red) is surrounded by three O (light blue) in a planar group, whereas Ca (dark blue) is octahedrally coordinated with six O.

to each other. Magnesite, siderite, rhodochrosite, and smithsonite, which all have a cation smaller than Ca and fitting the octahedral coordination, have the same structure as calcite. The situation is different in minerals of the aragonite group, which are discussed below.

Calcite, $CaCO_3$

Crystallography. Trigonal $\bar{3}2/m$; crystals common, exceptional wealth of forms, which are divided into three main types: (i) prismatic, short or long and terminated with {0001} or rhombohedron, (ii) rhombohedron only or as the prevalent form, and (iii) scalenohedron only or in combination with rhombohedron or prism; twinning common on {01$\bar{1}$2} and {0001}; also massive, granular, stalactitic, banded, oolitic or pisolitic, coral-shaped, earthy, etc.

Physical properties. Cleavage perfect on {10$\bar{1}$1}. Hardness 3; density 2.7. Mostly colourless or white but can be all colours owing to substitution or impurities; vitreous lustre; transparent to translucent. Very high birefringence.

Chemical properties, etc. The composition of calcite is usually close to its ideal formula, but Mn, Fe, and Mg can replace Ca to a limited extent. Calcite reacts readily with cold hydrochloric acid and forms H_2O and CO_2, with lively effervescence. Calcite is polymorphic with aragonite; its crystal structure is described above.

Names and varieties. *Iceland spar*, known from Iceland, is a particularly clear and transparent form of calcite demonstrating the marked birefringence. *Dog-tooth spar* is a variety characterized by crystals with steep scalenohedral terminations; *nail-head spar* has crystals with shallow terminations

Occurrence. Calcite is one of the most abundant minerals and occurs in many different associations. It is an important rock-forming mineral in sedimentary rocks and derived metamorphic rocks, and is almost the sole

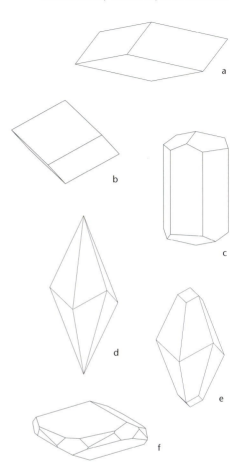

Figure 260. Calcite: (a) {01$\bar{1}$2}; (b) {10$\bar{1}$1}; (c) {10$\bar{1}$0} and {01$\bar{1}$2}; (d) {21$\bar{3}$1}, (e) {21$\bar{3}$1} and {10$\bar{1}$1}; (f) {10$\bar{1}$2}, {10$\bar{1}$0}, and {21$\bar{3}$1}.

mineral in rocks such as calcareous limestones and marbles. It is found in caves and cavities in these rocks as incrustations and stalactites, and is also formed as *calc-sinter* (*travertine*), etc. from springs. It occurs biogenetically in corals, shells, and the skeletons of a variety of marine animals. Calcite is found as a primary mineral in igneous rocks such as carbonatites and nepheline-syenites, as an alteration product of Ca-rich silicates, and infilling cavities of basalts in association

Figure 262. Calcite from Fakse, Denmark. Subject: 62 × 75 mm.

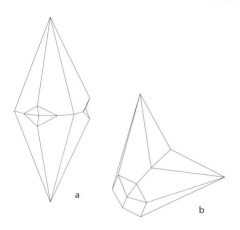

Figure 261. Twins of calcite: (a) on {0001}; (b) on {10$\bar{1}$1}; the crystal form in both (a) and (b) is the scalenohedron {21$\bar{3}$1}.

with zeolites. It is common in hydrothermal veins, generally in association with sulphides. Of the almost countless localities famous for fine calcite crystals, only a few can be mentioned, e.g. St. Andreasberg in the Harz Mountains, Germany, Joplin in Missouri, USA, and Dalnegorsk in Russia. Eskifjördhur in Iceland is noted for its flawless calcite.

Use. Calcite in the form of limestones, etc. is used for the production of cement, mortar, building stones, soil conditioners, and for many other purposes; it is accordingly among the principal raw materials. The highly transparent Iceland spar has been used for optical purposes, and was previously used for polarizing light.

Diagnostic features. Cleavage, birefringence, lively effervescence in cold dilute hydrochloric acid; and, in part, crystal habit and hardness.

Figure 263. Calcite, showing the three cleavage directions, from Iceland. Subject: 49 × 69 mm.

Figure 264. Calcite from Derbyshire, England. Subject: 36 × 65 mm.

Figure 266. Calcite surrounding a wire of silver, from Kongsberg, Norway. Crystal 20 mm across.

Figure 265. Calcite from Keweenaw Peninsula, Michigan, USA. Subject: 87 × 95 mm.

Figure 267. Magnesite with serpentine (green) from Nes, Snarum, Norway. Subject: 114 × 128 mm.

Magnesite, $MgCO_3$

Crystallography. Trigonal $\bar{3}2/m$; crystals uncommon, mostly rhombohedral; usually in dense earthy, chalk-like or porcelain-like masses, more rarely in granular or fibrous aggregates.

Physical properties. Cleavage perfect on {$10\bar{1}1$}. Hardness about 4; density 3.0. Colour is white, grey, yellowish, or brownish; vitreous lustre; transparent to translucent.

Chemical properties, etc. Mg can be replaced by Fe and in small amounts by Ca or Mn. Magnesite effervesces in warm hydrochloric acid. It has a calcite structure.

Occurrence. Magnesite occurs as veins and masses derived from the alteration of Mg-rich igneous rocks such as peridotites and serpentinites; magnesite of this type is usually cryptocrystalline and contains very fine-grained SiO_2. It is also found as bands or layers in chlorite- or talc-schists and in limestones, where it is either primary or an alteration product of calcite or dolomite; in the latter case magnesite is sometimes well crystallized.

Use. Magnesite is used for the production of bricks for furnace linings, magnesia cement, and, on a small scale, as a source of Mg.

Diagnostic features. Mode of occurrence, i.e. form of aggregate as well as mineral association; the flint-like forms are distinguished from flint, etc. by a lower hardness; effervescence in warm hydrochloric acid.

Figure 268. Siderite from Ivigtut, Greenland. Subject: 40 × 66 mm.

Siderite, FeCO₃

Crystallography. Trigonal $\bar{3}2/m$; crystals mostly {10$\bar{1}$1} or other rhombohedra, faces often curved or composite; occurs also granular, botryoidal, globular, oolitic or earthy.

Physical properties. Cleavage perfect on {10$\bar{1}$1}. Hardness 4; density 4.0. Colour is light to dark brown, reddish-brown, greyish or greenish; streak white; vitreous lustre; translucent.

Chemical properties etc. Mn and Mg can substitute for Fe; effervesces in warm hydrochloric acid. Siderite has a calcite structure.

Occurrence. Siderite is primarily found in connection with clays, slates, or coal-bearing sediments. In such deposits it is massive, fine-grained, or concretionary and common-

Figure 270. Rhodochrosite from Chalkidike, Greece. Subject: 60×85 mm.

ly occurs in mixtures with clays as *clay iron-stone* and mixed with clay or interstratified with coal as *black-band ore*. (The term *clay ironstone* is also used for mixtures of hematite and clay.) Siderite is also found in metamorphic rocks derived from such sediments. Siderite is a common gangue mineral in hydrothermal veins and is occasionally found in granitic and nepheline-syenite pegmatites. Siderite was an abundant mineral in the cryolite deposit at Ivigtut, Greenland. Fine siderite crystals are found in Mont Saint-Hilaire in Quebec, Canada, and in several localities in Minas Gerais, Brazil.

Use. Siderite is an Fe ore of local importance.

Diagnostic features. Colour and density as compared with other carbonates, and cleavage as compared with sphalerite and other minerals of similar colour; effervescence in warm hydrochloric acid.

Figure 269. Siderite from Belo Horizonte, Minas Gerais, Brazil. Field of view: 39×63 mm.

Rhodochrosite, $MnCO_3$

Crystallography. Trigonal $\bar{3}2/m$; crystals usually as $\{10\bar{1}1\}$ rhombohedron, more rarely as scalenohedron; faces sometimes curved or composite; mostly granular, botryoidal, or as incrustations.

Physical properties. Cleavage perfect on $\{10\bar{1}1\}$. Hardness $3\frac{1}{2}$–4; density 3.6. Colour is pink to deep red, also dark brown owing to presence of impurities; vitreous lustre; transparent to translucent.

Chemical properties, etc. Mn can be replaced by Fe, and to a lesser extent by Ca, Mg, or Zn; effervesces in warm hydrochloric acid. Rhodochrosite has a calcite structure.

Occurrence. Rhodochrosite occurs in hydrothermal veins, typically in association with Ag–Pb–Zn–Cu sulphides, other carbonates or Mn minerals such as manganite. It is also found in metasomatic manganese deposits together with, e.g. rhodonite, spessartine, and hausmannite, in pegmatites and as a sec-

191

ondary mineral in sedimentary manganese deposits. Particularly well-developed crystals are known from the Kalahari manganese field in the Republic of South Africa, Sweet Home mine and several other mines in Colorado, USA, and from the Huallapon mine and other places in Peru. Capillitas in Catamarca, Argentina, is known for a banded, stalactitic rhodochrosite that when polished is used for decorative objects.

Use. Rhodochrosite is an inferior Mn ore; it is sometimes used for decorative purposes.

Diagnostic features. Colour and cleavage; effervesces in warm hydrochloric acid; has a lower hardness than rhodonite.

Figure 271. Rhodochrosite from Peru. Subject: 20 x 32 mm.

Smithsonite, $ZnCO_3$

Crystallography. Trigonal $\bar{3}2/m$; crystals rare, most often rhombohedron or scalenohedron; occurs usually in reniform, stalactitic, or encrusting forms.

Physical properties. Cleavage perfect on $\{10\bar{1}1\}$. Hardness 4-4½; density 4.4. Colour mostly dirty brownish-white, also blue, green, yellow, pink, white, or colourless; streak white; strong vitreous lustre; translucent.

Chemical properties, etc. Fe and to a minor degree Mn replace Zn. Ca, Mg, Cd, Co, and Cu can also be present in small amounts; effervesces in warm hydrochloric acid. Smithsonite has a calcite structure.

Names and varieties. Smithsonite was formerly known as *calamine* in Britain; in the USA, hemimorphite was known as *calamine*.

Occurrence. Smithsonite is a supergene mineral found in the upper oxidized zones of Zn deposits, commonly as an alteration product of sphalerite. It typically occurs in association with cerussite, malachite, and Zn minerals such as hydrozincite and hemimorphite. Kelly near Magdalena in New Mexico, USA, Lavrion in Greece, Tsumeb in Namibia, and Broken Hill in New South Wales, Australia, are among the well-known localities.

Use. Smithsonite is locally of importance as a Zn ore. Smithsonite from a few localities, e.g. Lavrion, is used for ornamental purposes.

Diagnostic features. Density, effervescence in warm hydrochloric acid.

Dolomite, $CaMg(CO_3)_2$

Crystallography. Trigonal $\bar{3}$; crystals most often simple $\{10\bar{1}1\}$ rhombohedra, faces often curved and composite, giving crystals a saddle-like appearance; twinning on $\{0001\}$ common, also on $\{10\bar{1}0\}$, $\{11\bar{2}0\}$, and lamellar parallel to $\{02\bar{2}1\}$; mostly massive, coarse to fine granular, compact.

Physical properties. Cleavage perfect on $\{10\bar{1}1\}$. Hardness 3½-4; density about 2.9. Colour white or grey, also greenish, brown-

Figure 272. Smithsonite from Tsumeb, Namibia. Subject: 106 × 104 mm.

Figure 273. Smithsonite from Monte Masau, Sardinia, Italy. Subject: 63 × 120 mm.

Figure 274. Smithsonite from Tsumeb, Namibia. Subject: 71 × 85 mm.

193

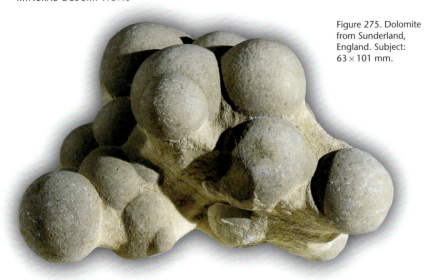

Figure 275. Dolomite from Sunderland, England. Subject: 63 × 101 mm.

ish, or pinkish; vitreous lustre; transparent to translucent.

Chemical properties, etc. Fe can replace Mg, and there is a complete solid-solution series between dolomite and ankerite. To a lesser extent Mn can replace Mg. Dolomite effervesces only slightly in cold hydrochloric acid but the reaction is lively if the acid is warm. The crystal structure of dolomite is a calcite structure with the modification that every

Figure 276. Dolomite from Navarra, Spain. Subject: 99 × 113 mm.

second Ca layer perpendicular to the *c* axis is replaced by an Mg layer. As Ca and Mg are ordered, the symmetry is reduced to $\bar{3}$ (compare with calcite).

Names and varieties. *Kutnahorite*, CaMn $(CO_3)_2$, has a dolomite structure with Mn instead of Mg; it is a rare mineral with properties like those of dolomite except for a white to pink colour.

Occurrence. Dolomite is widespread in sedimentary series and in the dolomitic marbles that are derived from metamorphism of such sediments. Dolomitic rocks are believed to have been deposited as calcitic and aragonitic sediments, and subsequently altered under the influence of Mg-rich solutions, either on the sea floor shortly after deposition or later by circulating groundwater. Dolomite is also found in hydrothermal veins, commonly in association with baryte, fluorite, and minerals containing Pb and Zn. Extensive occurrences of dolomitic rocks are found in the Dolomites of northern Italy; particularly well-developed crystals are known from, e.g. Navarra in Spain, Traversella in Italy, Lengenbach quarry in Binntal, Switzerland, and Tsumeb, Namibia.

Use. Dolomite is used as a building stone and in special types of cement.

Diagnostic features. Granular dolomite differs from calcite in showing lively effervescence only in warm hydrochloric acid. Crystals of dolomite are generally distinguished by crystal habit, cleavage, and, in comparison with minerals of the calcite group, by twin laws found only in $\bar{3}$.

Figure 277. Ankerite from Cornwall, England. Subject: 80×128 mm.

Ankerite, $CaFe(CO_3)_2$

Ankerite has a structure like that of dolomite and forms a complete solid-solution series with dolomite. Its physical properties closely resemble those of dolomite except that the colour is usually more brownish. The modes of occurrence are similar to those of dolomite; ankerite is in particular found in association with Fe-ore deposits, and also in Au-bearing quartz veins and the adjacent rocks.

The aragonite group

In the aragonite group the planar $[CO_3]$ groups are bonded to large cations, i.e. Ca^{2+} or the larger Ba^{2+}, Sr^{2+}, or Pb^{2+}. The cations are coordinated with nine O atoms in this structure, whereas the smaller cations in the calcite group are in sixfold coordination. Ca^{2+} is a medium-sized cation that can fit into both modes of coordination. The structure of aragonite is orthorhombic, pseudo-hexagonal. Its pseudo-hexagonal character is reflected in a tendency to form mimetic twins.

195

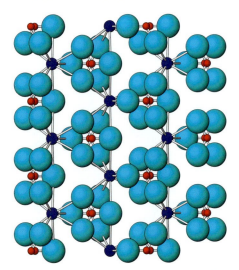

Figure 278. Part of the aragonite structure, viewed along the *c* axis. C (red) is surrounded by three O (light blue) in a planar group; Ca (dark blue) is coordinated with nine O.

Aragonite, CaCO₃

Crystallography. Orthorhombic $2/m2/m2/m$, pseudo-hexagonal; crystals common, typically long prismatic to needle-shaped with {110} prisms terminated by steep prisms, e.g. {091} and bipyramids, e.g. {991}, commonly in radiated aggregates, or short prismatic, slightly tabular parallel to {010} and with a flat {011} prism; twinning on {110} very common, often repeated to trillings simulating hexagonal prisms, or in thin polysynthetic lamellae seen as striations on {001}, both contact and penetration twins seen; also in columnar aggregates, as incrustations, stalactitic, coral-like, reniform, or pisolitic with radiating fibres.

Physical properties. Cleavage distinct on {010}, conchoidal fracture. Hardness 3½–4; density 2.9. Colourless, white, grey, yellowish, or with blue, green, violet, or red nuances; vitreous lustre; transparent to translucent.

Chemical properties, etc. Aragonite is polymorphic with calcite; its crystal structure is described above. It effervesces in cold hydrochloric acid.

Occurrence. Aragonite is less common than calcite and is more unstable than calcite under ordinary surface conditions; it is often replaced by calcite. It occurs as pisolitic or sinter-like deposits at hot springs, as stalactites in caves, in clays in association with gypsum, and with siderite in Fe-ore deposits, where it is sometimes found as coral-like formations known as *flos ferri*. Aragonite is also found in cavities in young basalts together with zeolites, in altered basic rocks with serpentines and in some marbles metamorphosed at low temperature and high pressure. The shells of some animals, such as the

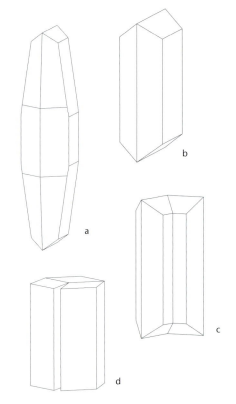

Figure 279. Aragonite: (a) {010}, {110}, {011}, {091}, and {991}; (b) {010}, {110}, and {011}; (c) twinned on {110}, both individuals have {010}, {110}, and {011}; (d) trilling on {110}; all three crystals have {110} and {001}.

Figure 280. Aragonite
from Sicily, Italy.
Subject: 46 × 62 mm.

pearly layer and pearls in some bivalves, consist entirely of aragonite, whereas most other shells are composed of calcite. Aragon in Spain and Agrigento (formerly Girgenti) in Sicily, Italy, are famous localities for twinned aragonite; Erzberg and Hüttenberg in Austria are known for spectacular flos ferri.

Diagnostic features. Lively effervescence in cold hydrochloric acid, as for calcite; distinguished from calcite by crystal habit, higher density, and a different cleavage.

Figure 281. Aragonite, pisolitic, from Karlory Vary (Carlsbad), Bohemia, Czech Republic.
Subject: 52 × 97 mm.

Figure 282. Strontianite from Strontian, Scotland. Subject: 56 × 65 mm.

Strontianite, $SrCO_3$

Crystallography. Orthorhombic $2/m2/m2/m$, pseudo-hexagonal; crystals uncommon, mostly long prismatic to needle-shaped; twinning on {110} common; usually columnar, granular or fibrous.

Physical properties. Cleavage good on {110}. Hardness 3½–4; density 3.8. Colour is white, grey, yellowish, or greenish; vitreous lustre; mostly translucent.

Chemical properties, etc. Ca can to some degree replace Sr. Strontianite has an aragonite structure. It effervesces with hydrochloric acid.

Occurrence. Strontianite occurs in low-temperature hydrothermal veins in limestones and marbles, commonly in association with baryte, calcite, and celestine, and sometimes with sulphides. It is also found as concretions in limestones and clays. Important deposits of strontianite are found, e.g. in marls near Hamm and Münster, Germany. Very fine strontianite is found at Strontian, the type locality.

Use. Together with celestine, strontianite is a source of Sr.

Diagnostic features. Density and (in contrast to celestine, commonly associated with it) effervescence in hydrochloric acid.

Witherite, $BaCO_3$

Crystallography. Orthorhombic $2/m2/m2/m$, pseudo-hexagonal; crystals uncommon, most often as repeated twins on {110} forming pseudo-hexagonal bipyramids with horizontally striated faces; also globular, botryoidal, granular, or coarsely fibrous.

Physical properties. Cleavage distinct on {010}. Hardness 3–3½; density 4.3. Colour is white, grey, or with light reflected colours; vitreous lustre; transparent to translucent.

Chemical properties etc. To a limited extent Ca and Sr can replace Ba. Witherite has an aragonite structure. It effervesces with hydrochloric acid. Witherite is poisonous.

Occurrence. Witherite is the second most widespread Ba mineral after baryte, but is much more rare; it occurs in hydrothermal veins, typically in association with galena, baryte, and fluorite. Fine crystals are known from, e.g. Cave-in-Rock, Illinois, USA, Hexham, Northumberland, and Alston Moor, Cumbria, in England.

Use. Witherite is an inferior ore of Ba.

Diagnostic features. Density and effervescence with hydrochloric acid.

Cerussite, $PbCO_3$

Crystallography. Orthorhombic $2/m2/m2/m$, pseudo-hexagonal; crystals common, crystal habit varies widely, e.g. tabular parallel to {010}, long prismatic parallel to the *a* axis, and with equal development of {111} and {021} resembling a hexagonal bipyramid; twinning on {110} and sometimes on {130} very common and often repeated to produce star-shaped groups of crystals or reticular formations; also granular or fibrous.

Physical properties. Cleavage distinct on {110}, indistinct on {021}. Hardness 3–3½; density 6.6. Colourless, white, or grey; adamantine lustre; transparent to translucent.

Chemical properties, etc. The composition of cerussite is normally close to the ideal for-

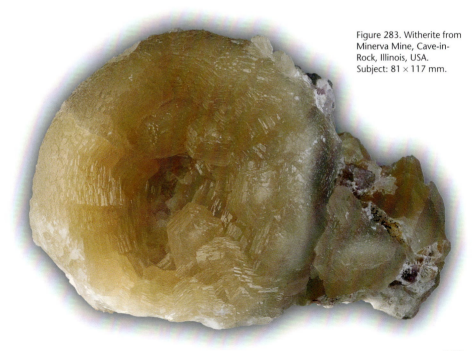

Figure 283. Witherite from Minerva Mine, Cave-in-Rock, Illinois, USA. Subject: 81 × 117 mm.

Figure 284. Cerussite from Bad Ems, Rheinland-Pfalz, Germany. Field of view: 31 × 43 mm.

mula. It effervesces weakly with cold hydrochloric acid and strongly with warm nitric acid. Cerussite has an aragonite structure.

Occurrence. Cerussite is a common supergene mineral formed by reaction between CO_2-containing solutions and galena; it typically occurs in association with primary minerals such as galena and sphalerite, and with secondary minerals like anglesite, pyromorphite, and smithsonite. Tsumeb in Namibia, Broken Hill in New South Wales, Australia, Monteponi in Sardinia, Italy, Flux mine, Arizona, and Wheatley mines, Pennsylvania, USA, are among the well-known localities for well-crystallized cerussite.

Use. Cerussite is an important ore of Pb.

Diagnostic features. Density, colour, lustre, and twinning.

Figure 285. Cerussite from Broken Hill, New South Wales, Australia. Field of view: 24 × 36 mm.

Figure 286. Malachite from Bisbee, Arizona, USA.
Field of view: 40 × 60 mm.

Malachite, $Cu_2CO_3(OH)_2$

Crystallography. Monoclinic $2/m$; good crystals rare, usually long prismatic to needle-shaped or hair-like and assembled in rosettes or bunches; twinning on {100} common; also seen as pseudomorphs after azurite and other minerals; mostly in globular, encrusting, botroyoidal, or stalactitic formations, commonly with a glossy surface and with concentric or parallel colour banding.

Physical properties. Cleavage perfect on {$\bar{2}01$}, distinct on {010}, uneven fracture in massive forms. Hardness $3\frac{1}{2}$–4; density 4.0, lower in massive forms. Colour is light to dark green; streak light green; strong vitreous lustre on crystal faces, otherwise silky lustre, sometimes dull; translucent.

Chemical properties, etc. Malachite effervesces with hydrochloric acid. Cu is present as Cu^{2+} and is accommodated in octahedra forming chains along the c axis; the chains are linked together by $[CO_3]$ groups.

Occurrence. Malachite is widespread and occurs as a secondary mineral in upper oxidized zones of Cu deposits, especially in association with limestones; it is typically found together with various Fe oxides and Cu minerals such as azurite, cuprite, copper, and chrysocolla. Among the best-known localities are the Russian occurrence near Nizhniy Tagil in the Urals, Russia, Tsumeb in Namibia, and Bisbee and other localities in Arizona, USA. In Europe malachite is especially known from Chessy near Lyon in France.

Use. Malachite is used in jewellery and decoration.

Diagnostic features. Colour, colour banding, mode of occurrence, and effervescence with hydrochloric acid.

Azurite, $Cu_3(CO_3)_2(OH)_2$

Crystallography. Monoclinic $2/m$; crystals common, habit varies, e.g. long prismatic parallel to the b or c axes or tabular parallel to {001}, usually with many faces; also massive, stalactitic, botryoidal, columnar, or radiating.
Physical properties. Cleavage perfect on {011}, distinct on {100}, conchoidal fracture.

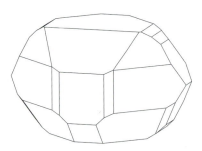

Figure 288. Azurite: {100}, {001}, {102}, {10$\bar{2}$}, {110}, {210}, {011}, {012}, {013}, {111}, {11$\bar{2}$}, and {12$\bar{3}$}.

Figure 287. Azurite from Touissit, Morocco. Subject: 33×72 mm.

Hardness 3½–4; density 3.8. Colour is azure-blue or deep blue, often lighter in massive forms; streak light blue; vitreous lustre; transparent to translucent.
Chemical properties, etc. Azurite effervesces in hydrochloric acid. Cu is present as Cu^{2+} and is accommodated in fourfold coordination in planar square groups forming chains along the b axis; the chains are linked together by $[CO_3]$ groups.
Occurrence. Azurite is less common than malachite but occurs in same way. Beautiful crystals have been found at Chessy near Lyon in France, Touissit in Morocco, Tsumeb in Namibia, and Bisbee in Arizona, USA.
Diagnostic features. Colour, mode of occurrence, and effervescence in hydrochloric acid.

Hydrozincite, $Zn_5(CO_3)_2(OH)_6$

Monoclinic; crystals usually tabular parallel to {100}; found mostly in earthy dense to porous masses and as banded incrustations or stalactites. Cleavage perfect on {100}; hardness 2–2½; density 3.2–3.8. Colour is pure white to grey or pale yellow; pearly lustre on crystal faces, otherwise earthy dull. Fluoresces bluish-white under ultraviolet light. Hydrozincite is a secondary mineral in the oxidized zones of ore deposits with sphalerite and other Zn ores.

Aurichalcite, $(Zn,Cu)_5(CO_3)_2(OH)_6$

Monoclinic; found as small crystals in velvet-like bunches but occurs mostly in incrustations like hydrozincite. Cleavage good in one direction; hardness 1–2; density about 4.2. Colour is light to dark green or sky-blue; silky lustre. Aurichalcite occurs in the upper oxidized zones of deposits containing Zn and Cu, usually associated with malachite, azurite, hemimorphite, or hydrozincite. It is known from, e.g. Lavrion, Greece.

Bastnäsite, $(Ce,La)CO_3F$

Hexagonal; crystals most often simple with {0001} and {10$\bar{1}$0} only; prism faces usually horizontally grooved. No distinct cleavage; hardness 4–4½; density about 5.0. Colour is yellowish-brown to reddish-brown; greasy vitreous lustre; mostly translucent. Bastnäsite is one of a group of closely related minerals distinguished as bastnäsite-(Ce), bastnäsite-(La), (La, Ce) CO_3F, and bastnäsite-(Y), (Y,Ce) CO_3F. Bastnäsite occurs in alkaline pegma-

Figure 289. Hydrozincite from Evelin mine, Pine Creek District, Northern Territory, Australia. Subject: 73×102 mm.

Figure 290. Aurichalcite from Kamarizos, Lavrion, Greece. Subject: 51×96 mm.

tites and hydrothermal veins as well as in contact or alteration zones. It is known, e.g. in association with an Fe deposit at Bastnäs in Sweden, from which it was first described. The rare-earth elements and Y are present in a number of other carbonates, usually with a similar crystal habit and mode of occurrence, e.g. *parisite*, $Ca(Ce,La)_2(CO_3)_3F_2$, and *synchysite*, $Ca(Ce,La)(CO_3)_3F$.

Leadhillite, $Pb_4(SO_4)(CO_3)_2(OH)_2$

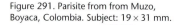

Monoclinic and usually found as well-formed crystals with a hexagonal outline and tabular parallel to {001}. Cleavage perfect on {001}; hardness 2½; density 6.5. Colourless, white,

Figure 291. Parisite from from Muzo, Boyaca, Colombia. Subject: 19×31 mm.

Figure 292. Phosgenite from Monteponi, Sardinia, Italy. Field of view: 34×41 mm.

grey, or pale coloured; resinous lustre, pearly lustre on {001}; transparent to translucent. Leadhillite is found as a supergene mineral in the oxidized zones of Pb deposits, commonly in association with cerussite and anglesite. It is known from a number of localities, e.g. the Susanna mine at Leadhills in Scotland.

Phosgenite, $Pb_2CO_3Cl_2$

Tetragonal and found as prismatic crystals, often dominated by {110} and {001} and with subordinate {111}. Cleavage perfect on {110} and {001}; hardness 2½-3; density 6.1. Colour is yellowish-white to yellow-brown or grey; resinous lustre; transparent to translucent. Phosgenite is a secondary mineral occurring in near-surface zones of Pb deposits as an alteration product of galena and other Pb minerals, often associated with cerussite. Some of the finest crystals are found at Monteponi and other localities in Sardinia, Italy.

Natron, $Na_2CO_3 \cdot 10H_2O$

Monoclinic; in nature occurs mostly as incrustations, coatings, or as an efflorescence. Hardness 1–1½; density 1.5. Colour is white, grey, or yellowish; vitreous lustre. Natron is readily soluble in water; in air, it transforms quickly to the orthorhombic *thermonatrite*, $Na_2CO_3 \cdot H_2O$. This mineral and the related *trona*, $Na_3(HCO_3)(CO_3) \cdot 2H_2O$, have the same appearance as soda. All three minerals are precipitated from salt lakes in arid regions; natron is usually precipitated in the cold period of the year and thermonatrite in the warm season.

Artinite, $Mg_2CO_3(OH)_2 \cdot 3H_2O$

Monoclinic; crystals needle-shaped or fibrous, assembled in pillows or spheres of radiating crystals. Cleavage perfect on {100}; hardness 2½; density 2.0. Colour is white with a silky lustre. Artinite is a low-temperature hydrothermal mineral found in veins in

Figure 293. Trona from Sweetwater Co., Wyoming, USA. Subject: 42 × 87 mm.

Figure 294. Artinite from San Benito County, California, USA. Subject: 15 × 21 mm.

205

Figure 295. Borax from Boron, California, USA. Subject: 36 × 50 mm.

Nitre, KNO_3, is orthorhombic with an aragonite structure. It occurs as thin incrustations or coatings, white or grey, with vitreous lustre. It is also readily soluble in water. Nitre is found in limestone cavities, e.g. in Hungary and Spain, and as an inferior mineral in nitratine deposits.

Borax, $Na_2B_4O_5(OH)_4 \cdot 8H_2O$

Crystallography. Monoclinic $2/m$; crystals prismatic, typically dominated by of {100} and {110}, more rarely {010} or {001}.

Physical properties. Cleavage perfect on {100}, distinct on {110}, conchoidal fracture; brittle. Hardness 2–2½; density 1.7. Colourless, alters by dehydration to white, grey, or yellowish; vitreous to dull lustre; translucent to opaque.

Chemical properties, etc. Borax dehydrates partly in dry air and alters to *tincalconite*, $Na_2B_4O_5(OH)_4 \cdot 3H_2O$. Borax is soluble in water and has a sweetish salty taste.

Occurrence. Borax occurs as an evaporate from salt lakes, usually in association with halite, gypsum, and various borates. It is found, e.g. at Borax Lake and other salt lakes in California and Nevada, USA. It is also known from Turkey, Argentina, and Tibet (where it was first known under the name *tincal*).

Use. Borax is one of the principal boron minerals. Borates are used for many purposes, e.g. for medical products, soaps and detergents, special glasses and enamels, in textiles, nonflammable materials, rocket fuel, and much more. Boron nitride is used for mortars and as an abrasive because of its hardness, which is comparable with that of diamond.

Diagnostic features. Crystal habit, mode of occurrence, density, and solubility.

altered ultrabasic rocks in association with, e.g. serpentine and brucite. It is known from a number of asbestos deposits, e.g. in Val Brutta, Italy.

Nitratine, $NaNO_3$

Trigonal with a calcite structure; crystals rhombohedral; in nature found only as granular masses. Cleavage as calcite; hardness 1½–2; density 2.2. Colourless or only light coloured, vitreous lustre. Readily soluble in water. Nitratine is primarily known from Chile, where it is found in a belt more than 600 km long in the deserts in the northern part. It occurs in layers alternating with layers of sand, halite, and gypsum; the deposit is mined as a source of N for fertilizers.

Ulexite, $NaCaB_5O_6(OH)_6 \cdot 5H_2O$

Triclinic; crystals needle-shaped, fibrous, or as capillaries, assembled in loose cotton-wool-like masses or in parallel packed aggregates with fibre optic properties. Hardness 2½, but barely measurable; density 2.0. Colour is white with a silky lustre. Ulexite occurs in the same ways as borax and is commonly associated with it.

Colemanite, $CaB_3O_4(OH)_3 \cdot H_2O$

Monoclinic; crystals normally short prismatic with many crystal forms; mostly in granular masses. Cleavage perfect on {010}, distinct on {001}; hardness 4–4½; density 2.4. Colourless, white, or pale yellowish; vitreous lustre; transparent to translucent. Colemanite occurs in borax deposits, e.g. in California, USA. It is largely found in cavities in sediments and is commonly associated with borax and ulexite. It has in the past been an economically important B mineral.

Kernite, $Na_2B_4O_6(OH)_2 \cdot 3H_2O$

Monoclinic; it occurs as large crystals or in metre-thick coarse-grained masses. Cleavage perfect on {100} and {001}; hardness 2½; density 1.9. Colourless, often with a white coating owing to alteration; vitreous lustre; transparent. Kernite is found in great quantities in the Kramer District in Kern Co., California, USA, where it occurs in clays that supposedly have been slightly contact metamorphosed. Today this occurrence is the economically most important B deposit in the USA.

Sassolite, $B(OH)_3$

Triclinic; occurs as small tabular pseudo-hexagonal crystals. Cleavage perfect on {001}; hardness 1, feels greasy; density 1.5. Colour is white to grey, pearly lustre; sour taste. Sassolite occurs as a sublimate at fumaroles and as a precipitate at hot springs, e.g. at Sasso in Tuscany, Italy.

Hambergite, $Be_2BO_3(OH)$

Orthorhombic; found as long prismatic well-formed crystals, faces often striated parallel to the c axis. Cleavage perfect on {010}; hardness 7½; density 2.4. Colour is greyish-white; vitreous lustre; transparent to translucent. Hambergite occurs in granitic and syenitic pegmatites and is known from, e.g. Little Three mine near Ramona, and Himalaya mine in California, USA.

Rhodizite, $(K,Cs)Be_4Al_4(B,Be)_{12}O_{28}$

Cubic; crystals mostly as rhombic dodecahedra or tetrahedra. No distinct cleavage; hardness > 8½; density 3.4. Colourless to white or

Figure 296. Colemanite from Boraxo Mine, Inyo County, California, USA. Field of view: 60 × 70 mm.

Figure 297. Rhodizite from Antsongombato, Mahaiza, Madagascar. Field of view: 35 × 56 mm.

yellow; strong vitreous lustre; mostly transparent. Rhodizite is found in granitic pegmatites and is known from various localities in the Urals, Russia, and in Madagascar.

Boracite, $Mg_3B_7O_{13}Cl$

Orthorhombic, cubic above 268 °C; crystals most often with cubic forms such as {100}, {110}, {111} or {1$\bar{1}$1}; also in dense, globular forms. No cleavage; hardness 7–7½; density 3.0. Colour is white to grey, also light greenish or bluish; vitreous lustre, at times dull; transparent to translucent. Strongly pyroelectric and piezoelectric. Boracite occurs in salt deposits and is especially known from the carnallite zone in the German Zechstein salt deposits, e.g. at Stassfurt and Lüneburg; it is known also from the similar Polish salt deposits at Inowrocław.

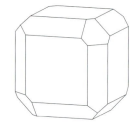

Figure 298.
Boracite: {100},
{110}, and {111}.

Figure 299. Gypsum from Zaragoza, Spain. Field of view: 40 × 50 mm.

Sulphates, chromates, molybdates, and tungstates

Sulphates are compounds of [SO$_4$] tetrahedra and one or more cations. Whereas sulphur is present as a large anion, S^{2-}, in sulphides, it is present as a very small cation, S^{6+}, in sulphates. The [SO$_4$] tetrahedron is a complex anion or group in which four O atoms are placed round a central S atom at the corners of a tetrahedron. These tetrahedra are isolated in the structure and do not form groups, chains, rings, or layers as in silicates. Similar conditions are found in chromates, molybdates, and tungstates.

Most sulphates form as weathering products in the upper oxidized zones of ore deposits or by evaporation from sea water or salt lakes. Baryte and a few other sulphates are mainly found as primary minerals in veins.

Thenardite, Na$_2$SO$_4$

Orthorhombic; crystals most often dominated by {111}; also as incrustations or an efflorescence. Cleavage perfect on {010}; hardness 2½–3; density 2.7. Colourless to white, grey, or yellowish; vitreous lustre. Readily soluble in water, tastes salty. Thenardite forms by evaporation from salt lakes in arid regions such as northern Chile, where it is usually found associated with other sulphates, nitrates, and carbonates. It is also found in vol-

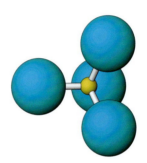

Figure 300. An [SO$_4$] tetrahedron with S (yellow) surrounded by four O (light blue). O is drawn at less than its true size in relation to S to reveal the structure.

Figure 301. Thenardite from Bolivia. Crystals about 15 mm high.

Figure 302. Glauberite from Villa Rubia, Ocana, Spain. Field of view: 60 × 67 mm.

canic regions as incrustations on lavas and around fumaroles. Extensive layers of thenardite are found near Aranjuez at Madrid, Spain, the locality from which it was first described.

Glauberite, $Na_2Ca(SO_4)_2$

Monoclinic; crystals variable, often with {001}, {100}, {110}, and {111}. Cleavage perfect on {001}; hardness 2½–3; density 2.8. Colourless or grey, yellowish or flesh-coloured; vitreous lustre. Sparingly soluble in water, tastes slightly salty. Glauberite is found in salt deposits, in cavities in basalts, at fumaroles, and with borates and nitrates in arid areas. Fine crystals are known from, e.g. Stassfurt, Germany.

Baryte, barite, $BaSO_4$

Crystallography. Orthorhombic $2/m2/m2/m$; crystals common, very diverse in habit and often face-rich, frequently tabular parallel to {001} encircled by a prism, e.g. {210}, or short to long prismatic parallel to the a, b, or c axes; crystals often assembled in rosettes or in crested forms resembling a mountain landscape; also granular.

Physical properties. Cleavage perfect on {001}, almost perfect on {210}, i.e. three directions with 90°, 90°, and 78° or 102° between them. Hardness 3–3½; density 4.5. Colourless, white, light blue, or green, also yellowish- to reddish-brown; vitreous lustre; transparent to translucent.

Chemical properties, etc. Sr can replace Ba, but baryte is generally close to its ideal for-

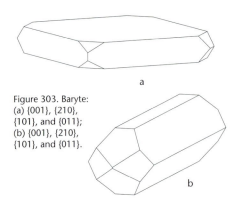

Figure 303. Baryte:
(a) {001}, {210},
{101}, and {011};
(b) {001}, {210},
{101}, and {011}.

mula. In baryte the large Ba cation is accommodated in twelvefold coordination.

Names and varieties. *Baryte* is the English spelling; *barite* is used in the USA.

Occurrence. Baryte is an abundant mineral and the most common Ba mineral. It occurs as a main mineral in hydrothermal veins, typically associated with fluorite, calcite, quartz, galena, and many other ore minerals. It is also common as fillings in veins and cavities in limestones, together with calcite, and can constitute essential parts of the residual clays formed by weathering of limestones. It occurs also as cement in sandstones. Among the many occurrences of well-crystallized baryte are Alston Moor and Frizington, Cumbria, and other places in England, Baia Sprie and other localities in Romania, and Elk Creek and elsewhere in the USA.

Use. Baryte has many applications, e.g. as drilling mud in oil and gas wells, as a filler in paper, as a pigment, in floor materials, etc., and as a source of Ba.

Diagnostic features. Density, cleavage, and crystal habit.

Celestine, $SrSO_4$

Crystallography. Orthorhombic $2/m2/m2/m$; crystals resemble those of baryte, commonly tabular parallel to {001}, also long prismatic parallel to one of the three axes; also granular and, rarely, fibrous.

Physical properties. Cleavage perfect on {001}, good on {210}. Hardness 3–3½; density 4.0. Colour is usually bluish-white, also colourless, greenish, or reddish; vitreous lustre; transparent to translucent.

Chemical properties, etc. Sr is normally in minor degree replaced by Ba. Celestine has a baryte crystal structure.

Occurrence. Celestine is not as widespread as baryte; it occurs dispersed in limestones and sandstones and in fissures and cavities in these rocks. It is also found as an inferior mineral in salt deposits and, less commonly, in hydrothermal veins. Beautiful crystals of celestine are found in association with sulphur in Sicily, Italy, and near Mahajanga (Majunga) in Madagascar.

Figure 304. Baryte from Egremont, Cumbria, England. Field of view: 46 × 60 mm.

Figure 305. Baryte from Rock Candy Mine, Grand Forks, British Columbia, Canada. Field of view: 29 × 46 mm.

Use. Celestine and strontianite are the principal sources of Sr. Sr compounds have various minor applications, e.g. in fireworks (for red colour).

Diagnostic features. Colour, cleavage, and density (which is slightly lower than baryte).

Anglesite, $PbSO_4$

Orthorhombic; crystal structure and crystal habit as in baryte, but crystals not as common; mostly granular, globular, or dense, also in concentric bandings round a core of galena. Cleavage as in baryte, but not as perfect, conchoidal fracture; hardness $2\frac{1}{2}$–3;

Figure 306. Celestine from Maradah, Libya. Subject: 67 × 89 mm.

213

Figure 307. Anglesite from Monteponi, Sardinia, Italy. Field of view: 32 × 31 mm.

density 6.3. Colourless, white to grey, yellowish, or greenish; resinous lustre on crystal faces; transparent to opaque. Anglesite occurs in the upper zones of ore deposits, largely as an alteration product of galena; it is usually associated with cerussite and gypsum. Particularly beautiful anglesite crystals are found at Monteponi in Italy, Touissit in Morocco, and Tsumeb in Namibia. The high density, resinous lustre, and association with galena are typical.

Anhydrite, $CaSO_4$

Crystallography. Orthorhombic $2/m2/m2/m$; crystals uncommon, mostly in massive or coarse crystalline masses, granular, or fibrous.

Physical properties. Cleavage perfect on {010}, almost perfect on {100}, and good on {001}, i.e. three directions perpendicular to each other. Hardness 3½; density 3.0. Colourless to grey or bluish; vitreous or pearly lustre; transparent to translucent.

Figure 308. Anhydrite from Siglo X Mine, Naica, Chihuahua, Mexico. Subject: 80 × 97 mm.

Chemical properties, etc. Ba or Sr is usually present in small amounts. Anhydrite has a different crystal structure from that of baryte; in anhydrite, Ca is in eightfold coordination.
Occurrence. Anhydrite has the same mode of occurrence as gypsum but is less widespread. It is found in salt deposits, where it occurs in layers alternating with layers of gypsum, halite, and limestone. Anhydrite occurs in some hydrothermal veins. Large deposits of anhydrite are found in the area around Stassfurt in Germany and at Halle in Austria. Particularly beautiful lavendercoloured crystals are known from the Simplon tunnel in Switzerland.
Use. Anhydrite is used, e.g. as a soil conditioner, but is not as important as gypsum.
Diagnostic features. Cleavage and, especially in comparison with gypsum, lustre and hardness.

Gypsum, $CaSO_4 \cdot 2H_2O$

Crystallography. Monoclinic $2/m$; crystals common, usually simple, typically tabular parallel to{010} bordered by {120} and {11$\bar{1}$}, faces sometimes curved and crystals bent or twisted; contact and penetration twins common on {100} (proper swallowtail-twins), also twins on {$\bar{1}$01}; also granular, dense, fibrous, or concretionary.
Physical properties. Cleavage perfect on {010}, distinct on {100} and {011}; cleavage folia flexible but inelastic. Hardness 2; density 2.3. Colourless; also white, grey, yellowish, or brownish; vitreous lustre, pearly lustre on cleavage surfaces, also silky lustre when fibrous; transparent to translucent.

215

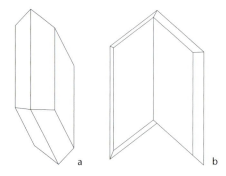

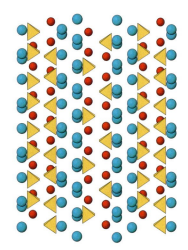

Figure 309. Gypsum: (a) {010}, {120}, and {11$\bar{1}$}; (b) 'swallow-tail' twin on {100} with the same forms as in (a).

Chemical properties, etc. Gypsum is normally close to its ideal formula. It has a layered crystal structure, in which layers of [SO$_4$] tetrahedra and Ca alternate with layers of H$_2$O; the layers are parallel to {010}. Ca is bonded to six O from [SO$_4$] tetrahedra and two H$_2$O. The bonds within the H$_2$O layers

Figure 310. A slice of the crystal structure of gypsum seen with the c axis N–S and the b axis E–W. Gypsum has a layered structure with layers of [SO$_4$] tetrahedra (yellow) and Ca (red) alternating with layers of H$_2$O (blue); the layers are parallel to {010}. Ca is bonded to six O from [SO$_4$] tetrahedra and to two H$_2$O. The bonds within the H$_2$O layers are weak, which explains the perfect cleavage after {010}.

Figure 311. Gypsum from Gorgue, Murcia, Spain. Field of view: 45 × 79 mm.

Figure 312. Gypsum var. satin spar from Kungur, Urals, Russia. Subject: 92 × 103 mm.

are weak, which is reflected in the perfect cleavage on {010}.

Names and varieties. *Selenite* is sometimes used for clear well-crystallized gypsum; *alabaster* for massive, fine-grained, and, occasionally, for banded gypsum; and *satin spar* for fibrous gypsum.

Occurrence. Gypsum is a common mineral, especially in sedimentary rocks, often as extensive beds associated with limestones, clays, halite, and other salt minerals. Series of this kind are formed by evaporation of sea water. As gypsum is sparingly soluble, it is among the first salts to be precipitated, followed by anhydrite, halite, and eventually the readily soluble Mg- and K-containing salts. Anhydrite can take up water and transform to gypsum, and as this process is accompanied by an increase in volume, such deposits are often strongly folded. Gypsum is the main constituent of the cap rocks on salt domes. It also occurs as a precipitate from salt lakes, at fumaroles, as lenses and concretions, as an efflorescence, as dispersed crystals in chalk and clays, and as a secondary mineral in ore deposits. There are countless localities for well-crystallized gypsum; some of the more spectacular crystals come from Fuentes de Ebro at Zaragoza in Spain, Naica and Santa Eulalia in Chihuahua, Mexico, Mt. Elliott, Queensland, and Walaroo, South Australia, Australia, and Hanksville, Utah, USA.

Use. Gypsum is used for wallboards and similar products for building construction, as a soil conditioner, in cement, and as a filler in paper and paints.

Diagnostic features. Hardness, cleavage, and crystal habit, including twinning.

Chalcanthite, $CuSO_4 \cdot 5H_2O$

Triclinic; natural crystals rare; occurs mostly as stalactites, incrustations or coatings. No distinct cleavage, conchoidal fracture; hardness 2½; density 2.3. Colour is blue; vitreous lustre; transparent. Chalcanthite, which is

Figure 313. Epsomite from Calatayud, Aragon, Spain. Subject: 66 × 105 mm.

soluble, is a common alteration product from chalcopyrite and other Cu ore minerals; it is of some importance as a Cu ore, e.g. in Chuquicamata, Chile.

Melanterite, $FeSO_4 \cdot 7H_2O$

Monoclinic; natural crystals rare; mostly stalactitic, as incrustations or coatings, often with a fibrous structure. Cleavage perfect on {001}, distinct on {110}; hardness 2; density 1.9. Shades of green, bluish when Cu replaces Fe; vitreous lustre; mostly translucent; readily soluble in water. Melanterite forms by the disintegration of pyrite, marcasite, and other Fe sulphides and is typically found in the upper oxidized zones of ore bodies; it is very common on mine walls.

Epsomite, $MgSO_4 \cdot 7H_2O$

Orthorhombic, 222; natural crystals rare; found especially as fibrous incrustations or coatings. Cleavage perfect on {010}, distinct on {101}; hardness 2–2½; density 1.7. Colourless or white; vitreous lustre, silky in fibrous forms; mostly translucent; readily soluble in water, has a bitter metallic taste. Epsomite is found in the upper altered parts of Mg-rich rocks and in limestone cavities and galleries, commonly together with pyrite; it is also found at springs and salt lakes. It is known from, e.g. Epsom in Surrey, England.

Brochantite, $Cu_4SO_4(OH)_6$

Monoclinic; crystals uncommon, most often prismatic or needle-shaped in loose aggregates; occurs mostly as incrustations or in dense, granular masses. Cleavage perfect on {100}; hardness 3½–4; density 4.0. Emerald-green to greenish-black; vitreous lustre; mostly translucent. Brochantite is found in the oxidation zone in Cu deposits in arid regions, commonly together with malachite and chrysocolla. It can be difficult to distinguish from atakamite, antlerite, and other Cu sulphates. *Antlerite*, $Cu_3SO_4(OH)_4$, is an orthorhombic mineral with the same appearance and mode of occurrence as brochantite.

Halotrichite, $FeAl_2(SO_4)_4 \cdot 22H_2O$

Monoclinic; crystals needle-shaped or hair-like, commonly assembled in fibrous aggregates. No distinct cleavage; hardness 1½; density 1.9. Greyish-white, yellowish, or greenish; silky lustre; readily soluble in water. Halotrichite occurs as a weathering product of pyrite in Al-rich rocks; it is also found as an efflorescence in mines, especially coal mines, and at fumaroles.

Figure 314. Brochantite from Tsumeb, Namibia. Subject: 66 × 72 mm.

Figure 315. Halotrichite from Golden Queen Mine, Mojave, California, USA. Field of view: 66 × 99 mm.

Figure 316. Linarite from Grube Ortiz, Sierra Capillitas, Argentina. Subject: 80×94 mm.

Polyhalite, $K_2Ca_2Mg(SO_4)_4 \cdot 2H_2O$

Triclinic; crystals rare, most often long prismatic; found mostly massive or fibrous. Cleavage perfect on $\{10\bar{1}\}$; hardness 3–3½; density 2.8. Colourless, white, or with pale borrowed colours; vitreous lustre, somewhat greasy. Polyhalite is widespread in many marine salt deposits, commonly in association with halite and anhydrite; it is known from, e.g. Stassfurt, Germany.

Mirabilite, $Na_2SO_4 \cdot 10H_2O$

Monoclinic; natural crystals rare, found mostly as an efflorescence, fibrous coatings, or stalactitic. Cleavage perfect on $\{100\}$; hardness 1½; density 1.5. Colourless to white; vitreous lustre; readily soluble in water. Mirabilite occurs at salt lakes and springs. It is known from, e.g. the Great Salt Lake in Utah, USA, where it is precipitated during the winter when the solubility drops. *Glauber salt* is an old name for mirabilite.

Linarite, $CuPbSO_4(OH)_2$

Monoclinic; crystals often long prismatic parallel to the *b* axis, face-rich; also found as encrustations. Cleavage perfect on $\{100\}$; hardness 2½; density 5.4. Colour is azure-blue with a light blue streak; strong vitreous lustre. Linarite occurs in the oxidation zone of ores containing Pb and Cu. It is known from, e.g. Linares, Jaén, in Spain, and (as beautiful crystals) from Red Gill mine and other mines in Cumbria, England.

Alunite, $KAl_3(SO_4)_2(OH)_6$

Crystallography. Trigonal *3m*; crystals uncommon, most often in cube-like rhombohedra, rarely tabular parallel to {0001}; mostly massive, dense, granular, or earthy, commonly mixed with other minerals such as quartz and kaolinite.

Physical properties. Cleavage distinct on {0001}. Hardness 3½–4; density 2.8. Colour is white, yellowish, or reddish; vitreous lustre, pearly on {0001}; mostly translucent.

Chemical properties, etc. Na usually replaces K and can be predominant.

Names and varieties. *Jarosite*, $KFe_3(SO_4)_2$·$(OH)_6$, is an Fe analogue of alunite. It is yellow to brown and is widespread as coatings on pyrite and other Fe ores, from which it is formed.

Occurrence. Alunite is common in the near-surface zone of rocks affected by ascending sulphuric acid solutions of volcanic origin. Such 'alunitized' rocks are found, e.g. at Tolfa near Rome, Italy, where alunite has been mined for generations for the production of alum and as an ore of aluminium.

Diagnostic features. None; larger masses of alunite can resemble limestones.

Kieserite, $MgSO_4$·H_2O

Monoclinic; crystals rare, mostly granular. Cleavage perfect on {110} and {111}; hardness 3½; density 2.6. Colourless, grey, or yellowish; vitreous lustre. Kieserite occurs in marine salt deposits, usually with polyhalite, anhydrite, and halite, and is found in larger quantities in, e.g. northern German and Russian salt deposits.

Figure 317. Jarosite from Chihuahua, Mexico. Subject: 81 × 110 mm.

Figure 318. Kainite from Wilhelmshall, Germany. Subject: 65 × 92 mm.

Kainite, $KMg(SO_4)Cl\cdot 3H_2O$

Monoclinic; crystals rare, tabular; mostly granular. Cleavage perfect on {001}; hardness 3½; density 2.1. Colourless or with pale borrowed colours; vitreous lustre. Kainite is found in marine salt deposits and occurs in larger amounts in the north German potash deposits, e.g. at Stassfurt, and is found together with sylvite, halite, and carnallite.

Hanksite, $KNa_{22}(SO_4)_9(CO_3)_2Cl$

Hexagonal; found as well-developed prismatic crystals, often large. Cleavage good on {0001}; hardness 3–3½; density 2.6. Colourless, grey, or pale yellowish, vitreous lustre, somewhat greasy; mostly translucent. Hanksite occurs in the B-rich salt lakes in California, USA, together with borax, trona, and halite.

Figure 319. Hanksite from Boron, California, USA. Subject: 69 × 110 mm.

Ettringite, $Ca_6Al_2(SO_4)_3(OH)_{12}\cdot26H_2O$

Hexagonal; crystals usually needle-shaped. Cleavage perfect on $\{10\bar{1}0\}$; hardness 2–2½; density 1.8. Colourless to white or yellowish; silky lustre. Ettringite occurs in cavities in limestone inclusions in basaltic lavas, e.g. at Ettringen in Eifel, Germany. It is also found in well-developed crystals up to 20 cm long in the Kuruman Mn field in the Republic of South Africa.

Crocoite, $PbCrO_4$

Crystallography. Monoclinic $2/m$; crystals typically long prismatic parallel to the c axis and with striated faces, often partially hollow; also massive, columnar, or granular.

Physical properties. Cleavage distinct on $\{110\}$. Hardness 2½–3; density 6.0. Colour is red or orange-red; streak orange-yellow; strong vitreous lustre; transparent to translucent.

Occurrence. Crocoite is a secondary mineral occurring in the upper alteration zone of ore deposits, especially in Pb-rich veins cutting Cr-bearing ultramafic rocks; it is generally found in association with anglesite, pyromorphite, cerussite, and similar secondary Pb minerals. The best-known occurrence of crocoite is at Dundas in Tasmania, Australia, but fine crocoite crystals are also found at Yekaterinburg in the Urals, Russia.

Diagnostic features. Colour, lustre, and density.

Wolframite, $(Fe,Mn)WO_4$

Crystallography. Monoclinic $2/m$; crystals normally prismatic and slightly tabular parallel to $\{100\}$, faces striated parallel to the c-axis, crystals often in subparallel aggregates; also granular or lamellar.

Physical properties. Cleavage perfect on $\{010\}$. Hardness about 5; density 7.1–7.5, in-

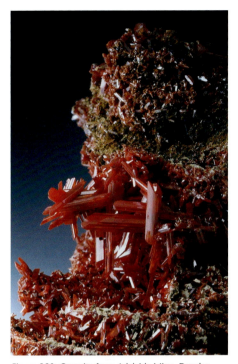

Figure 320. Crocoite from Adelaide Mine, Dundas, Tasmania, Australia. Field of view: 60×90 mm.

creasing with Fe content. Colour is brownish-black to black; streak reddish-brown to black; submetallic lustre; almost opaque.

Chemical properties, etc. Wolframite encompasses a complete solid-solution series between *ferberite*, $FeWO_4$, and *hübnerite*, $MnWO_4$; wolframite usually has an intermediate composition.

Occurrence. Wolframite occurs in pneumatolytic and granite pegmatitic veins, where it is typically found associated with quartz, cassiterite, topaz, tourmaline, arsenopyrite, and Li micas. It is also known from hydrothermal veins, together with sulphides such as pyrite and pyrrhotite. Particularly fine crystals are known from, e.g. Panasqueira in Portugal, Pasto Bueno in Peru, Tong Wha in Korea, and from Boulder and other counties in Colora-

223

Figure 321. Wolframite from Tong Wha, Korea. Subject: 20 × 42 mm.

do, USA. Wolframite was found in small quantities in the cryolite deposit at Ivigtut, Greenland.

Use. Wolframite is the principal ore of W. W is primarily used in hard steel alloys, abrasives, and for filaments in electric light bulbs.

Diagnostic features. Colour, lustre, cleavage, and density.

Scheelite, $CaWO_4$

Crystallography. Tetragonal $4/m$; crystals common, mostly with bipyramids {101} or {112}; penetration twins on {110} common; also granular.

Physical properties. Cleavage distinct on {101}. Hardness 4½–5; density 6.1. Colour is white, yellow to brown, also greenish or reddish; greasy lustre, often resinous; mostly translucent.

Chemical properties, etc. W can be replaced by Mo, but generally only in small amounts.

Figure 322. Ferberite from Mundo Nuevo, La Libertad, Peru. Subject: 105 × 145 mm.

Figure 323. Hübnerite from Pasto Bueno, Ancash, Peru. Subject: 42 × 75 mm.

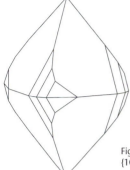

Figure 324. Scheelite: {101}, {112}, {211}, and {123}.

Names and varieties. *Powellite*, $CaMoO_4$, has a scheelite structure and forms an incomplete solid-solution series with scheelite. Spectacular crystals of powellite are found with apophyllite and zeolites in cavities in basalt near Nasik, India.

Occurrence. Scheelite occurs in the same types of pegmatitic and hydrothermal veins as wolframite, often in association with it. It is also found in contact-metamorphic rocks, especially where granites have intruded limestones, and is here associated with grossular, wollastonite, epidote, and other Ca-bearing silicates; occurrences of this type are economically the most important. There are numerous localities with well-crystallized scheelite, especially in China, Korea, and Brazil; in Europe, e.g. Traversella in Italy, Cínovec (Zinnwald) in the Czech Republic, and Kammegg in Switzerland; and in the USA, e.g. the Boriana and Cohen mines in Arizona.

Use. Scheelite is an important ore of W.

Diagnostic features. Crystal habit, colour, lustre, and density.

Figure 325. Scheelite from Xinjiang, China. Field of view: 28 × 31 mm.

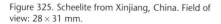

225

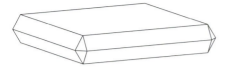

Figure 326. Wulfenite: {001}, {112}, and {101}.

Wulfenite, $PbMoO_4$

Crystallography. Tetragonal $4/m$; crystals common, normally tabular parallel to {001} encircled by small faces of pyramids or prisms; also very thin with practically only {001}; also granular.

Physical properties. Cleavage distinct on {101}. Hardness 3; density 6.8, decreasing with increasing substitution of Ca. Colour is yellow, orange-yellow, orange-red, white, or grey; adamantine to resinous lustre; transparent to translucent.

Chemical properties, etc. W and Ca can to some degree replace Mo and Pb.

Occurrence. Wulfenite is a secondary mineral found in the upper oxidized zones of ore deposits with Pb and Mo; it is typically associated with pyromorphite, vanadinite, and cerussite. Beautiful crystals are found in many localities, e.g. in the Glove, Red Cloud, and Mammoth-St. Anthony mines in Arizona, USA.

Use. Wulfenite is an Mo ore of minor importance.

Diagnostic features. Crystal habit, colour, lustre, and mode of occurrence.

Figure 327. Wulfenite from Bou Becker, Morocco. Field of view: 22 × 30 mm.

Figure 328. Wulfenite from Red Cloud Mine, La Paz County, Arizona, USA. Field of view: 30 × 41 mm.

Phosphates, arsenates, and vanadates

Figure 329. Pyromorphite from Bunker Hill mine, Kellogg, Idaho, USA. Subject: 73 × 126 mm.

Phosphates are compounds of a [PO₄] tetrahedron and one or more cations. A [PO₄] tetrahedron consists of a phosphorus atom surrounded by four oxygen atoms located at the corners of a tetrahedron. The [PO₄] tetrahedron is generally isolated in the crystal structure.

Similar conditions are valid for arsenates and vanadates, which have [AsO₄] tetrahedra or [VO₄] tetrahedra respectively. Phosphorus (P), arsenic (As), and vanadium (V) replace each other quite extensively in these minerals, and in consequence the chemistry of the group can be complicated.

Some phosphates are primary minerals in igneous rocks and in pegmatites, whereas others are secondary minerals formed by weathering processes under near-surface conditions.

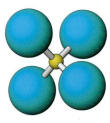

Figure 330. A [PO₄] tetrahedron with P (yellow) surrounded by four O (light blue). O is drawn at less than its true size in relation to P in order to reveal the structure.

Figure 331. Purpurite from Hohenstein, Erongo, Namibia. Subject: 65 × 83 mm.

Triphylite, LiFePO₄

Orthorhombic; crystals rare, mostly coarse granular. Cleavage good on {001}, less good on {010}; hardness 4½–5; density 3.6. Colour is bluish- to greenish-grey; vitreous lustre, somewhat resinous; translucent. Triphylite forms a complete solid-solution series with *lithiophilite*, $LiMnPO_4$, which is brown or salmon-coloured. Both occur in granite pegmatites, typically in association with other phosphates, spodumene, and beryl. Triphylite is known from many localities, e.g. the Newport and Palermo mines in New Hampshire, USA, and Hagendorf in Bayern, Germany. *Heterosite*, $FePO_4$, and *purpurite*, $MnPO_4$, are closely related minerals, found as oxidation products on surfaces of triphylite and lithiophilite, as coarse-grained masses, and also in pegmatites.

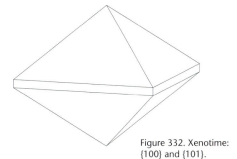

Figure 332. Xenotime: {100} and {101}.

Colour is yellowish- to reddish-brown; resinous lustre; translucent. Radioactive.

Chemical properties, etc. The ratios between the three rare-earth elements Ce, La, and Nd vary; each can be predominant, but Ce is the most common. Th is almost always present, up to 10–15%, which explains the radioactivity of monazite.

Xenotime, YPO₄

Tetragonal; crystals mostly small with crystal forms as zircon; also tabular parallel to {001}. Cleavage on {100}; hardness 4–5; density 4.5–5.1. Colour is yellowish, brownish, or greyish; greasy lustre; transparent to opaque. Y can to some degree be replaced by rare-earth elements, U, or Th. Xenotime occurs in muscovite-rich pegmatites, commonly associated with zircon, and occasionally in crystallographically orientated intergrowth with zircon. It is found, e.g. at Hidra in Norway and Ytterby in Sweden.

Monazite, (Ce,La,Nd,Th)PO₄

Crystallography. Monoclinic $2/m$; crystals uncommon, usually short prismatic and somewhat tabular parallel to {100}, faces commonly uneven and striated; twinning on {100} common; mostly granular, also as sand.
Physical properties. Cleavage good on {100}, less good on {010}. Hardness 5–5½; density 4.6–5.4, increasing with the Th content.

Figure 333. Monazite from Iveland, Norway. Subject: 18 × 22 mm.

229

Occurrence. Monazite occurs as an accessory mineral in granites, syenites, and associated pegmatites. It is also found in sands in association with other resistant minerals such as magnetite, ilmenite, and zircon. Well-crystallized monazite is known from, e.g. Mars Hill in North Carolina, USA, and from several pegmatites in Madagascar.

Use. Monazite is mined for its content of Ce and Th. The principal deposits are beach sands in Brazil, India, and Australia.

Diagnostic features. Crystal habit, density, and hardness.

Herderite, $CaBePO_4(F,OH)$

Monoclinic; crystals prismatic with many prisms. Cleavage on {110}; hardness 5–5½; density 3.0. Colourless or pale yellowish; vitreous lustre; transparent to translucent. Herderite is found in granitic pegmatites.

Figure 334. Herderite from Linopolis, Minas Gerais, Brazil. Subject: 31 × 63 mm.

Figure 335. Amblygonite from Montebras, Creuse, France. Subject: 83 × 96 mm.

Amblygonite, $(Li,Na)AlPO_4(F,OH)$

Triclinic; well-developed crystals uncommon, occurs mostly in coarse-grained masses. Cleavage perfect on {100}, good to distinct on {110} and {0$\bar{1}$1}; hardness 5½–6; density 3.0. Colour is white or pale yellow, green, blue, or red; vitreous lustre, somewhat greasy, pearly lustre on cleavage surfaces; transparent to translucent. The composition varies widely, especially in Li/Na and F/OH ratios; in *montebrasite*, $LiAlPO_4(OH,F)$, OH > F. Amblygonite and montebrasite occur in granitic pegmatites, frequently in metre-sized crystals, and typically with apatite and other Li-bearing minerals such as spodumene and lepidolite. They resemble feldspars, but are distinguished from them by cleavage and density. Exceptional crystals are known from Newry in Maine, USA.

Triplite, $Mn_2PO_4(F,OH)$

Monoclinic; crystals rare, occurs mostly in coarse-grained masses. Cleavage good on {001}, less good on {010}; hardness 5–5½; density 3.5–3.9. Colour is light brown, flesh-coloured, to nearly black; greasy lustre; translucent. Mn can be replaced by Fe or Mg, and there are solid-solution series between triplite and *zwieselite*, Fe_2PO_4 (F,OH), and triplite and *wagnerite*, $Mg_2PO_4(F,OH)$. All three minerals occur in granitic pegmatites, typically in association with apatite and other phosphate minerals. Triplite is known from, e.g. Branchville in Connecticut, USA.

Libethenite, $Cu_2PO_4(OH)$

Orthorhombic; crystals mostly short prismatic and octahedron-like owing to equal development of {110} and {011}. Cleavage indistinct on {100} and {010}; hardness 4; density 4.0. Colour is light to dark green or greenish-black; vitreous lustre; translucent. Libethen-

Figure 336. Libethenite from Rokana Open Pit, Zambia. Field of view: 27 × 41 mm.

ite occurs in the upper oxidized zones of Cu deposits, usually associated with malachite and azurite.

Olivenite, $Cu_2AsO_4(OH)$

Orthorhombic; crystal habit varies, short to long prismatic, needle-shaped or tabular; also fibrous, granular, earthy. Cleavage indistinct on {011} and {110}; hardness 3; density 3.9-4.5. Colour is olive-green or brownish-green; strong lustre; translucent to opaque. Olivenite occurs in the oxidation zone of As-rich Cu deposits, e.g. in Cornwall.

Figure 337. Olivenite from Broken Hill, Zambia. Subject: 37 × 41 mm.

Figure 338. Adamite from Mapimi, Durango, Mexico. Field of view: 26 × 39 mm.

Figure 339. Adamite from Lavrion, Greece. Field of view: 30 × 45 mm.

Adamite, $Zn_2AsO_4(OH)$

Orthorhombic; crystals face-rich and long prismatic parallel to the b axis, often in fan-shaped rosettes or radiated aggregates. Cleavage good on {101}; hardness 3½; density 4.4. Colour is honey-yellow or brown, pale green, white, or colourless; vitreous lustre; transparent to translucent. Adamite is a secondary mineral occurring in the oxidation zone of As-rich Zn deposits. Particularly beautiful crystal groups are found at, e.g. Tsumeb in Namibia and Ojuela in Mapimi, Mexico.

Lazulite, $(Mg,Fe)Al_2(PO_4)_2(OH)_2$

Monoclinic; crystals tapering pyramidal with {111} and {$\bar{1}$11}; also granular. No marked cleavage; hardness 5–6; density 3.1. Colour is sky-blue, greenish-blue, or bluish-white; vit-reous lustre; transparent to translucent. Mg can be replaced by Fe and there is a solid-solution series between lazulite and *scorzalite*, $(Fe,Mg)Al_2(PO_4)_2(OH)_2$. Lazulite occurs in strongly metamorphosed rocks and is known from, e.g. Champion mine, California, Graves Mountains, Georgia, and the Yukon, USA.

Pseudomalachite, $Cu_5(PO_4)_2(OH)_4$

Monoclinic; crystals rare; mostly fibrous, bo-tryoidal, massive, banded. No distinct cleav-age; hardness 4½–5; density 4.3. Colour is emerald-green to greenish-black, fibrous forms often lighter coloured; vitreous lustre; translucent. Pseudomalachite occurs in the oxidation zone in Cu deposits, usually asso-ciated with malachite and azurite.

233

Figure 340. Lazulite from Rapid Creek, Yukon, Canada. Subject: 41 × 71 mm.

Figure 341. Pseudomalachite from Broken Hill, Zambia. Subject: 108 × 106 mm.

Figure 342. Descloizite from Abenab, Namibia. Subject: 56 × 90 mm.

Descloizite, PbZnVO$_4$(OH)

Orthorhombic; crystals commonly prismatic dominated by {110} prism and {111} bipyramid, assembled in radiating or subparallel aggregates; also fibrous, granular, or stalactitic. No cleavage; hardness 3½; density 6.2. Colour is brown to brownish-black or reddish-brown; greasy lustre; translucent. Zn can be replaced by Cu, and there is a solid-solution series between descloizite and *mottramite*, PbCuVO$_4$(OH). Descloizite is a secondary mineral found in the oxidation zone of ore deposits, commonly associated with vanadinite and pyromorphite. It is also found crystallized in sandstones; particularly fine crystal groups of this kind are known from Abenab and other localities in Namibia.

Brazilianite, NaAl$_3$(PO$_4$)$_2$(OH)$_4$

Monoclinic; crystals prismatic. Cleavage good on {010}; hardness 5½; density 3.0. The colour is yellow or greenish-yellow; vitreous lustre; transparent. Brazilianite occurs in cavities in granitic pegmatites. It is known especially from the Brazilian occurrences in Minas Gerais, where gemstone-quality brazilianite is found.

Figure 343. Brazilianite from Mendes Pimentel, Minas Gerais, Brazil. Field of view: 28 × 32 mm.

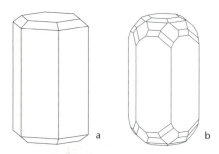

Figure 344. Apatite: (a) {10$\bar{1}$0}, {10$\bar{1}$1}, and {0001}; (b) {10$\bar{1}$0}, {0001}, {10$\bar{1}$1}, {10$\bar{1}$2}, {11$\bar{2}$1}, {20$\bar{2}$1}, and {21$\bar{3}$1}.

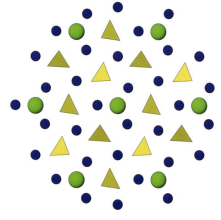

Figure 345. A slice of the crystal structure of apatite seen along the c axis. The fourth corner of the [PO$_4$] tetrahedra (yellow) cannot be seen in this view. Ca (blue) is located in two different positions, one in nine-fold coordination, the other in eightfold coordination. The large green anions are F, Cl, or OH. Note that the ions are not drawn in their correct proportions and that the [PO$_4$] tetrahedra are drawn schematically without regard to the size of the ions involved.

Apatite, $Ca_5(PO_4)_3F$

Crystallography. Hexagonal $6/m$; crystals common, tabular parallel to {0001} or short to long prismatic with {10$\bar{1}$0}, {10$\bar{1}$1}, and

Figure 346. Apatite from Oksøykollen, Snarum, Norway. Field of view: 78×127 mm.

Figure 347. Apatite from Renfrew County, Ontario, Canada. Subject: 51 × 79 mm.

tion series between these three end-members. In *carbonate-apatites* the [PO$_4$] group is partially replaced by a [CO$_3$OH] group. In addition, P can to some extent be replaced by Si, which has to be coupled with another substitution in order to maintain charge balance. Finally, to a lesser extent Mn or Sr can replace Ca. In the crystal structure of apatite Ca is accommodated in two different positions, one in ninefold and the other in eightfold coordination.

Names and varieties. *Britholite*, (Ce,Ca,)$_5$-(SiO$_4$,PO$_4$)$_3$(OH,F), is a brown mineral related to apatite; it is found in alkaline rocks and associated pegmatites, e.g. in the Ilímaussaq complex, Greenland.

Occurrence. Apatite is the principal phosphate mineral and is widespread as an accessory mineral in rocks of all types; it is also found in pegmatites and in veins. Apatite occurs in large amounts in Ti-containing magnetite deposits and in association with alkaline complexes. The largest known occur-

minor {0001}, other prisms and bipyramids occur; also granular, compact.

Physical properties. No distinct cleavage, conchoidal or uneven fracture. Hardness 5; density 3.2. Colour is yellowish-, greyish-, or bluish-green, brown, or colourless, often dirty; greasy vitreous lustre; transparent to translucent.

Chemical properties, etc. Apatite includes three mineral species: *fluorapatite*, Ca$_5$(PO$_4$)$_3$F, the most common; *hydroxylapatite*, Ca$_5$(PO$_4$)$_3$(OH), and *chlorapatite*, Ca$_5$(PO$_4$)$_3$Cl. There are complete solid-solu-

Figure 348. Apatite from Untersulzbachtal, Austria. Field of view: 32 × 38 mm.

237

Figure 349. Phosphorite, a phosphate rock consisting principally of collophane, from Sombrero, Leward Islands, West Indies. Subject: 95 × 116 mm.

rence of apatite is at Kirovsk in the Kola Peninsula, Russia, where a huge apatite body is situated between two nepheline-syenite intrusions. Apatite is also found as large masses in the magnetite deposit at Kiruna in Sweden and with perovskite and magnetite in the Gardiner complex, Greenland. Large crystals of apatite, up to 200 kg, are found at Renfrew and other localities in Ontario, Canada. Exceptional crystals are known from Mount Apatite in Main, USA, and from several mines in Durango, Mexico.

Apatite, primarily as carbonate-apatite, is furthermore widespread as the main constituent of *collophane*, which occurs as fine-grained or cryptocrystalline masses, usually as concretions or banded layers. It is formed by the accumulation of bones and teeth from animals, by reaction between guano and limestones, or by chemical precipitation from sea water.

Use. Apatite is the principal phosphate ore mineral; the majority of mined apatite is used as fertilizer.

Diagnostic features. Crystal habit, colour, lustre, and hardness.

Pyromorphite, $Pb_5(PO_4)_3Cl$

Crystallography. Hexagonal $6/m$; crystals common, mostly simple prismatic with $\{10\bar{1}0\}$ and $\{0001\}$, perhaps also $\{10\bar{1}1\}$, sometimes barrel-shaped and partially hollow; also globular.

Physical properties. Cleavage indistinct on $\{10\bar{1}1\}$, uneven fracture. Hardness $3\frac{1}{2}$–4; density 7.0. Colour is yellow, brown, or green; resinous lustre; mostly translucent.

Chemical properties, etc. P can be replaced by As, and there is a complete solid-solution series between pyromorphite and mimetite, $Pb_5(AsO_4)_3Cl$. Pyromorphite has an apatite structure.

Occurrence. Pyromorphite is a secondary mineral occurring in the upper oxidized parts of Pb deposits, generally in association with other secondary Pb minerals. Beautiful crystal groups are found at several places, e.g. at Ems in Germany, Bunker Hill in Idaho, USA, Caldbeck Fells, Cumbria, Wanlockhead, Dumfries and Galloway, and Leadhills, South Lanarkshire.

Use. Pyromorphite is a Pb ore of minor importance.

Diagnostic features. Crystal habit, density, and lustre.

Mimetite, $Pb_5(AsO_4)_3Cl$

Hexagonal with structure as pyromorphite; crystals mostly simple, barrel-shaped with $\{10\bar{1}0\}$ and $\{0001\}$; also globular. Cleavage indistinct on $\{10\bar{1}1\}$, conchoidal to uneven fracture; hardness $3\frac{1}{2}$–4; density 7.3. Colour is light yellow to yellowish-brown or orange-yellow; resinous lustre; mostly translucent. As can be replaced by P, and there is a complete solid-solution series between mimetite and pyromorphite. The mode of occurrence for mimetite is the same as for pyromorphite but mimetite is less widespread. Particularly beautiful crystals are found at Tsumeb, Namibia. The variety *campylite*, with barrel-shaped crystals, is found at Dry Gill, Cumbria, England.

Figure 351.
Mimetite from
Tsumeb, Namibia.
Subject: 28 × 47 mm.

Figure 350. Pyromorphite from Broken Hill, New South Wales, Australia. Field of view: 36 × 54 mm.

Figure 352. Mimetite from San Pedro Corralitos, Chihuahua, Mexico. Field of view: 76 × 120 mm.

Figure 353. Vanadinite from Mibladen-Aouli, Morocco. Field of view: 26 × 39 mm.

Vanadinite, $Pb_5(VO_4)_3Cl$

Crystallography. Hexagonal $6/m$; crystals common, mostly prismatic with $\{10\bar{1}0\}$ and $\{0001\}$ or tabular parallel to $\{0001\}$, sometimes hollow; also globular or as encrustations.

Physical properties. No cleavage, uneven fracture. Hardness 3; density 6.9. Colour is red, orange-red or reddish-brown; nearly adamantine lustre; transparent to translucent.

Chemical properties, etc. P and As can to a small extent replace V, and Ca, Zn, Cu, replace Pb. Vanadinite has an apatite structure.

Occurrence. Vanadinite is a relatively rare mineral found in the upper oxidized parts of Pb deposits. Among the well-known localities for vanadinite are Mibladen in Morocco, and the Old Yuma mine and other mines in Arizona and New Mexico, USA.

Figure 355. Variscite from Fairfield, Utah, USA. Subject: 126×139 mm.

Use. Vanadinite is an inferior ore of V and Pb.

Diagnostic features. Crystal habit, colour, density, and lustre.

Variscite, $AlPO_4 \cdot 2H_2O$

Orthorhombic; found mostly in cryptocrystalline forms. Hardness 4½, density 2.5; apple-green colour with a wax-like lustre. Variscite occurs as precipitations from circulating water in near-surface cavities; it is associated with a series of exotic phosphates. The best-known occurrence of variscite is near Fairfield in Utah, USA, where it is mined for ornamental purposes. *Strengite*, $FePO_4 \cdot 2H_2O$, has the same structure as variscite, but differs in appearance; it is found as spherical aggregates of red radiating fibres formed by alteration of triphylite and other pegmatitic phosphates. *Scorodite*, $FeAsO_4 \cdot 2H_2O$, also has

Figure 354. Vanadinite from Morocco. Field of view: 18×33 mm.

Figure 356. Scorodite from Berezovskiy, Urals, Russia. Subject: 60 × 68 mm.

the same structure as variscite; it occurs as small green crystals or coatings on arsenopyrite and other As-bearing minerals, from which it is formed.

Vivianite, $Fe_3(PO_4)_2 \cdot 8H_2O$

Crystallography. Monoclinic $2/m$; crystals common, mostly prismatic with predominant {010} and {100}; also as concretions, fibrous or bladed coatings, or earthy.

Physical properties. Cleavage perfect on {010}, cleavage folia flexible. Hardness 1½–2; density 2.7. Colourless when fresh, darkens quickly in air to blue or green to almost black owing to partial oxidation of Fe^{2+} to Fe^{3+}; vitreous lustre; transparent to translucent.

Chemical properties, etc. Mn, Mg, and Ca can in small amounts replace Fe.

Occurrence. Vivianite occurs as an alteration product of iron ores or Fe-bearing phosphates under near-surface conditions. It is also found as coatings and concretions in young sediments containing organic material such as bones, etc. Splendid crystals are known from the upper parts of tin mines in Bolivia.

Diagnostic features. Hardness, colour properties, and flexible cleavage folia.

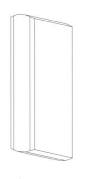

Figure 357. Vivianite: {100}, {010}, {110}, {10$\bar{1}$}, and {11$\bar{1}$}.

Erythrite, $Co_3(AsO_4)_2 \cdot 8H_2O$

Monoclinic; crystals rare, mostly small, needle-shaped; usually as powdered coatings. Cleavage perfect on {010}; hardness 1–2½; density 3.1. Colour is crimson-red, streak lighter red; vitreous lustre. It occurs as an alteration product of skutterudite and other Co-bearing ores and is normally found as coatings on these minerals. The apple-green *annabergite*, $Ni_3(AsO_4)_2 \cdot 8H_2O$, occurs similarly as coatings on Ni-bearing ores.

Figure 358. Vivianite from Morococala Mine, Ururo, Pantaleon Dalence Province, Ururo, Bolivia. Field of view: 90 × 106 mm.

Figure 359. Erythrite from Bou Azzer, Morocco. Subject: 76 × 78 mm.

Figure 360. Annabergite from Lavrion, Greece. Subject: 61 × 78 mm.

243

Figure 361.
Scholzite from Reaphook
Hill, South Australia, Australia.
Subject: 82 × 143 mm.

Scholzite, $CaZn_2(PO_4)_2 \cdot 2H_2O$

Orthorhombic; well-developed crystals common, typically needle-shaped. No cleavage; hardness 4; density 3.1. Colourless with a vitreous lustre; transparent. Scholzite occurs as an alteration product of phosphates in pegmatites and in sediments, e.g. at Reaphook Hill in South Australia, Australia.

Struvite, $(NH_4)MgPO_4 \cdot 6H_2O$

Orthorhombic *mm*2; crystals distinctly hemimorphic, showing different crystal forms at the two ends of the *c* axis (see Figure 362). Cleavage good on {001}; hardness 1½–2; density 1.7. Colour is yellowish- or brownish-

white; vitreous lustre; transparent to translucent when fresh, but becomes dull and chalky in dry air owing to dehydration. Struvite forms in deposits of bird or bat guano, in sediments, and in soils by bacterial activity, and in the human urinary system as kidney or bladder stones. It is known from, e.g. a postglacial marine deposit at Ålborg, Limfjorden in Denmark and from a peaty deposit under St. Nikolai Church, Hamburg in Germany.

Wavellite, $Al_3(PO_4)_2(OH,F)_3 \cdot 5H_2O$

Orthorhombic; single crystals rare, mostly in radiating spherical or hemispherical aggregates. Cleavage good on {110} and {101}; hardness 3½–4; density 2.4. Colourless, grey, yellowish, or greenish with a vitreous lustre; translucent. Wavellite occurs as a secondary mineral in fissures and cavities in slates and phosphate deposits, as in Highdown quarry, Filleigh, Devon (the type locality) and in Montgomery and other counties in Arkansas, USA.

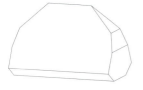

Figure 362.
Struvite: {00$\bar{1}$},
{010}, {012},
{021}, {101},
and {10$\bar{3}$}.

Figure 363. Struvite from Nicolai Kirsche, Hamburg, Germany. Field of view: 24 × 36 mm.

Figure 364. Wavellite from Montgomery County, Arkansas, USA. Field of view: 70 × 105 mm.

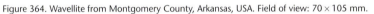

Figure 365. Turquoise from Los Cerrillos Mountains, Santa Fe County, New Mexico, USA. Subject: 31 × 49 mm.

Turquoise, $CuAl_6(PO_4)_4(OH)_8 \cdot 4H_2O$

Triclinic; crystals very rare, occurs mostly in dense, fine-grained, or cryptocrystalline masses or incrustations. Slightly conchoidal fracture, brittle; hardness 5–6; density 2.6–2.8. Colour is blue, bluish-green, or green; wax-like lustre, dull, translucent. Turquoise is a secondary mineral formed in thin veins and fissures in altered volcanic rocks in arid regions;

Figure 366. Turquoise from Santa Fe County, New Mexico, USA. Subject: 64 × 93 mm.

Figure 367. Autunite from Margnac, Haute-Vienne, France. Subject: 64 x 92 mm.

it is commonly associated with limonite and chalcedony. Among the most famous occurrences of turquoise are the Iranian in a trachyte at Neyshabur (Nishapur) in Khorasan, and the American, also in an altered trachyte, in Los Cerrillos Mountains in New Mexico. In the granites of the St. Austell area of Cornwall it is found as a secondary alteration product of metalliferous hydrothermal mineralization. Turquoise has from the earliest times been appreciated as a gemstone, preferably cut in various round forms. Turquoise is commonly cut by thin veins of other minerals. Such material is sold as *turquoise matrix*, but much of what is on sale is an imitation.

Autunite, $Ca(UO_2)_2(PO_4)_2 \cdot 10H_2O$

Tetragonal; crystals thin and tabular parallel to {001}, commonly in fans, scaly aggregates,

Figure 368. Torbernite from Bergen, Vogtland, Germany. Field of view: 18×27 mm.

or in uneven crusted coatings. Cleavage perfect on {001}; hardness 2–2½; density 3.1–3.2. Lemon-coloured to light green; vitreous lustre, pearly on cleavage surfaces; translucent; radioactive. Autunite is a secondary mineral formed in the oxidation zone of deposits of uraninite and other U-bearing minerals. *Torbernite*, $Cu(UO_2)_2(PO_4)_2 \cdot 10H_2O$, is a related mineral with the same structure, crystal habit, and physical properties as autunite except for the green colour; it has the same mode of occurrence as autunite.

Carnotite, $K_2(UO_2)_2(VO_4)_2 \cdot 3H_2O$

Monoclinic; crystals rare, mostly as powder or loose aggregates. Cleavage perfect on {001}; hardness about 2; density 4–5. Colour is light yellow to greenish-yellow; earthy, dull; radioactive. Carnotite is a secondary mineral formed in the weathering zone of U deposits. It is widespread as dispersed grains in sandstones in Colorado and Utah, USA, where it is an important ore of U and, to a lesser extent, an ore of V.

Figure 369. Quartz (var. smoky quartz) with orthoclase (whitish), aegirine (green), and eudidymite (small light-coloured plates), all silicates. Zomba–Malosa complex, Chilwa Province, Malawi. Field of view: 52 × 78 mm.

Silicates

The silicates constitute more than 90% of the earth's crust and are accordingly by far the largest mineral group. The feldspar group alone constitutes about 60% of the crust, and quartz slightly more than 10%. The silicates are the predominant minerals in igneous rocks and in most metamorphic and sedimentary rocks. Whereas a relatively small number of silicates are very common, a large number are more or less rare.

The fundamental structure unit in all silicates is the [SiO_4] tetrahedron, in which a central silicon atom (Si) is surrounded by four oxygen atoms (O) in a tetrahedral arrangement. The bond between silicon and oxygen is very strong, and is intermediate between an ionic and a covalent bond. The [SiO_4] tetrahedra can be isolated or in various groupings. Silicates are classified according to these arrangements (see Figure 371).

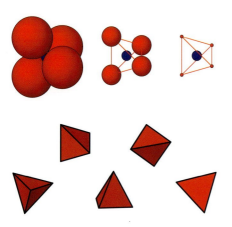

Figure 370. The [SiO_4] tetrahedron is the fundamental structure unit in all silicates. It consists of a central Si atom (blue) surrounded by four O atoms (red). In the upper left-hand corner the atoms are drawn in approximately their correct proportions. The oxygen atoms hide the central Si atom. In order to reveal the structure, the relative sizes of the atoms can be changed while retaining their true positions (top, centre). The ultimate solution is to simplify the drawing by removing the atoms altogether, showing only the lines connecting the O atoms. The tetrahedron is now evident. In the lower part of the figure a tetrahedron is seen from various angles.

Nesosilicates. The [SiO_4] tetrahedra are isolated, i.e. not linked to other tetrahedra (*nesos*, Gk., 'island'). The silicate part of the formula is in the simple case SiO_4, as e.g. in olivine, $(Mg,Fe)_2SiO_4$.

Sorosilicates. The [SiO_4] tetrahedra are linked to form pairs by sharing one corner (*soros*, Gk., 'group'). The silicate part of the formula is in the simple case Si_2O_7, as in hemimorphite, $Zn_4Si_2O_7(OH)_2 \cdot H_2O$.

Cyclosilicates. The [SiO_4] tetrahedra are linked to form rings, in which each tetrahedron shares corners with two others (*cyclo*, Gk., 'ring'). The silicate part of the formula is Si_nO_{3n}, where n is the number of tetrahedra in the ring, e.g. in beryl, $Be_3Al_2Si_6O_{18}$. Other kinds of rings exist.

Inosilicates. The [SiO_4] tetrahedra form infinite chains (*ino*, Gk., 'chain'), e.g. single chains, in which each tetrahedron shares corners with two others. The silicate part of the formula is in this case Si_2O_6, as in diopside, $CaMgSi_2O_6$. The tetrahedra can also form double chains, as in amphiboles, where [SiO_4]-tetrahedra alternately share two and three corners with others.

Phyllosilicates. All the [SiO_4] tetrahedra share corners with three others, thus forming infinite layers (*phyllo*, Gk., 'leaf'). The silicate part of the formula is Si_2O_5, e.g. in kaolinite, $Al_2Si_2O_5(OH)_4$.

Tectosilicates. All [SiO_4] tetrahedra share all four corners with others to form a three-dimensional framework (*tecto*, Gk., 'framework'). The silicate part of the formula is SiO_2, as in quartz, SiO_2, or in orthoclase, $KAlSi_3O_8$ (Al replacing one Si).

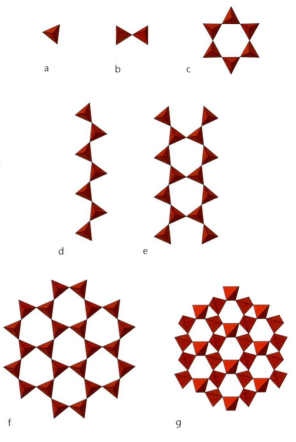

Figure 371. The classification of silicates is based on the number of corners the $[SiO_4]$ tetrahedra share with other $[SiO_4]$ tetrahedra. In nesosilicates (a) no corners are shared; in sorosilicates (b) one corner is shared; in cyclosilicates (c) two corners are shared to form rings, here a sixfold ring; in inosilicates two corners are shared to form infinite chains (d) or alternatively two and three corners to form infinite double chains (e); in phyllosilicates (f) three corners are shared; in tectosilicates (g) all four corners are shared. In (g) a three-dimensional lattice is projected down on the plane of the paper; the other structures are, by nature, approximately two-dimensional.

The Si/O ratio gradually changes through the groups from 1/4 to ½ as the polymerization of $[SiO_4]$ tetrahedra increases.

Aluminium (Al) is an important element in some silicates, as in tectosilicates, which would all have the same formula, SiO_2, if Al did not partially replace Si. Al is trivalent and Si tetravalent; so when Al^{3+} replaces Si^{4+} in the tetrahedra, compensation of some kind has to take place elsewhere in the structure. This typically takes place by 'introducing' a monovalent ion such as K^+ into suitable cavities in the structure or by replacing one K^+ with one Ca^{2+}. The classic example of such a coupled substitution is found in the feldspars, in which $Al^{3+} + Ca^{2+}$ can replace $Si^{4+} + K^+$ and vice versa, the balance being maintained.

Other elements can be accommodated in the spaces between the $[SiO_4]$ tetrahedra. Cat-ions such as Mg^{2+}, Fe^{2+}, Fe^{3+}, Mn^{2+}, Al^{3+}, and Ti^{4+} can enter these positions in octahedral coordination, i.e. surrounded by six oxygen atoms. Even though they have different va-lencies, they substitute readily, and compen-

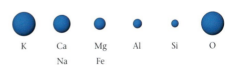

Figure 372. The most common elements in silicates are oxygen (O), silicon (Si), aluminium (Al), iron (Fe), calcium (Ca), sodium (Na), potassium (K), and magnesium (Mg). They occur as charged atoms, i.e. negative anions or positive cations, and are found as O^{2-}, Si^{4+}, Al^{3+}, Fe^{2+} or Fe^{3+}, Ca^{2+}, Na^+, K^+, and Mg^{2+}, shown here in approximately their correct proportions.

251

sation is made in other positions. The same applies when, e.g. Na^+ and Ca^{2+} replace each other. These cations are larger and fit into positions that are typically surrounded by eight oxygen atoms.

The size of the Al^{3+} ion makes it able to enter either a tetrahedral or an octahedral position, and this property is to a considerable degree responsible for the diversity of the silicates.

Figure 373. Andradite var. melanite in natrolite from the Gardiner complex, Greenland. Field of view: 60×69 mm.

Nesosilicates

In the nesosilicates the $[SiO_4]$ tetrahedra are isolated and are linked together by other cations in six- or eightfold coordination, more rarely in fourfold or fivefold coordination. Olivine and the garnets belong to this group. The general formulae for these minerals end with SiO_4 or a multiple of SiO_4.

Figure 374. Phenakite in quartz from Tangen, Kragerø, Norway. Field of view: 50×72 mm.

Phenakite, Be_2SiO_4

Crystallography. Trigonal $\bar{3}$; crystals prismatic or tabular rhombohedral, usually well-developed with several faces; twinning on $\{10\bar{1}0\}$ frequent.

Physical properties. Cleavage indistinct on $\{10\bar{1}0\}$; conchoidal fracture. Hardness $7\frac{1}{2}$–8; density 3.0. Colourless or whitish; vitreous lustre; transparent or translucent.

Chemical properties, etc. Phenakite is isostructural with willemite.

Occurrence. Phenakite is a rare mineral found in pegmatites, typically in association with topaz or beryl. Clear perfect crystals are found at San Miguel di Piracicaba and other places in Brazil; large crystals, up to about 20 cm long, are found at Kragerø, Norway.

Use. Phenakite is sometimes used as a gemstone.

Diagnostic features. Crystal habit, differs from quartz by a greater hardness and by not having striated prism faces.

Willemite, Zn_2SiO_4

Crystallography. Trigonal $\bar{3}$; crystals rare, mostly massive, granular.

Physical properties. Cleavage good on $\{0001\}$, uneven to conchoidal fracture. Hardness $5\frac{1}{2}$; density about 4. Colour is white or pale yellow, greenish, brownish, or reddish, rarely black; vitreous to greasy lustre; mostly translucent. Mn-bearing willemite fluoresces yellow-green in ultraviolet light.

Chemical properties, etc. Willemite is isostructural with phenakite. Fe is usually present in small amounts, whereas Mn can replace Zn to a considerable degree.

Names and varieties. *Troostite* is an Mn-bearing variety.

Occurrence. Willemite is found in crystalline limestones and to a lesser extent in the oxidized zones of Zn ore deposits. Franklin, New Jersey, USA, is the principal site of occurrence of willemite, which is here found in association with, e.g. franklinite and zincite.

Use. At Franklin, New Jersey, willemite was a valuable Zn ore.

Diagnostic features. Willemite usually fluoresces in UV light, but is otherwise difficult to identify. Specimens from Franklin are best identified by the associated minerals.

Figure 375. Willemite var. troostite, from Franklin, New Jersey, USA. Field of view: 60 × 72 mm.

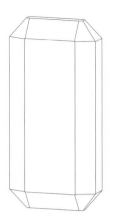

Figure 376. Olivine: {100}, {010}, {001}, {110}, {101}, {021}, and {111}.

Names and varieties. Olivine's name reflects its common olive-green colour. *Peridot* is olivine of gemstone quality. *Chrysolite* was previously a synonym for peridot, but is now occasionally used for olivines with 70–90% forsterite, *Tephroite*, Mn_2SiO_4, is a member of the olivine group. *Monticellite*, $CaMgSiO_4$, was formerly grouped with olivine; it is colourless, whitish, or yellowish, and occurs in contact-metamorphic limestones.

Occurrence. Olivine is an important rock-forming mineral. It typically occurs in me-

Olivine, $(Mg,Fe)_2SiO_4$

Crystallography. Orthorhombic $2/m2/m2/m$, crystals usually combinations of pinacoids, prisms, and bipyramids; mostly granular.

Physical properties. No distinct cleavage; conchoidal fracture. Hardness 6½–7; density 3.3 (forsterite) to 4.4 (fayalite). Colour is from yellowish-green, olive-green, and brown to black with increasing Fe content; vitreous lustre; transparent to translucent.

Chemical properties, etc. There is a complete solid-solution series between *forsterite*, Mg_2SiO_4, and *fayalite*, Fe_2SiO_4. Fe is present as Fe^{2+} and substitutes directly for Mg^{2+}. Common olivine has Mg > Fe. In the crystal structure of olivine, layers of $[(Mg,Fe)O_6]$ octahedra are linked by isolated $[SiO_4]$ tetrahedra. Mg and Fe are randomly distributed in the octahedral positions.

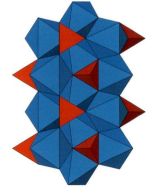

Figure 377. The crystal structure of olivine consists of layers of $[(Mg,Fe)O_6]$ octahedra (blue) sharing edges. These layers are linked together by isolated $[SiO_4]$ tetrahedra (red). Mg and Fe are randomly distributed in the octahedral positions.

lanocratic (dark-coloured) basic or ultrabasic igneous rocks such as basalt, peridotite, and dunite. In dunite, olivine is by far predominant. It alters to antigorite or other serpentine minerals. Forsterite is also found in metamorphosed dolomitic limestones. Olivine is a common mineral in stony meteorites and constitutes a substantial part of the upper mantle of the earth. Well-developed, transparent crystals are known from Zebirget (St. Johns Island), Red Sea, Egypt.

Use. Mg-rich olivine has a high melting point and is useful for refractory bricks; it is also used for insulation material.

Diagnostic features. Crystal habit, glassy appearance with respect to colour and fracture, granular mode. Mg-rich olivine does not occur associated with quartz.

Figure 378. Olivine from Zebirget (St. John's Island), Red Sea, Egypt. Subject: 16 × 20 mm.

Figure 379. Olivine from Almklovdalen, Norway. Subject: 73 × 87 mm.

Figure 380. Chondrodite in calcite from Pargas, Finland. Field of view: 26 × 39 mm.

The humite group

The humite group includes a series of closely related Mg silicates. Like olivine, they have layers of $[(Mg,Fe)O_6]$ octahedra linked by isolated $[SiO_4]$ tetrahedra. They differ from each other and from olivine in the way in which the octahedra are arranged within the layer.

Crystallography. Orthorhombic $2/m2/m2/m$ or monoclinic $2/m$; crystals usually with many forms, but mostly found granular; twinning common.

Physical properties. No distinct cleavage; conchoidal fracture. Hardness 6–6½; density about 3.2. Colour is yellowish to dark brown or dark red, also whitish; vitreous lustre; transparent to translucent.

Chemical properties, etc. The group includes *norbergite*, $Mg_3(SiO_4)(F,OH)_2$; *chondrodite*, $(Mg,Fe,Ti)_5(SiO_4)_2(F,OH,O)_2$; *humite*, $(Mg,Fe)_7(SiO_4)_3(F,OH)_2$, and *clinohumite*, $(Mg,Fe,Ti)_9(SiO_4)_4(F,OH)_2$. These formulae illustrate that Fe can to a limited extent replace Mg, that in chondrodite and clinohumite Ti can partially replace Mg, and that F and OH substitute for each other.

Names and varieties. A similar series of Mn-containing minerals exists, e.g. *alleghanyite*, $Mn_5(SiO_4)_2(OH)_2$.

Occurrence. These minerals occur mainly in contact-metamorphic dolomitic limestones.

Diagnostic features. Mode of occurrence and, to some extent, colour are the best diagnostic characteristics, but these minerals are generally difficult to identify and especially to distinguish from each other.

Garnets, $A_3B_2(SiO_4)_3$

In the general formula, A is Mg, Fe, Mn, or Ca; B is Al, Fe, or Cr.

Garnets are widespread minerals and are particularly abundant in metamorphic rocks. They are usually easily recognizable, since they are generally found as well-developed crystals with characteristic cubic crystal forms. Garnets can be difficult to distinguish from each other without chemical analysis, but may be identified approximately by mode of occurrence and colour. The name *garnet* is derived from the Latin word meaning 'grain'.

Figure 381. Titaniferous clinohumite (red) with magnetite (black) and richterite (light yellow) from the Gardiner complex, Greenland. Field of view: 66 × 100 mm.

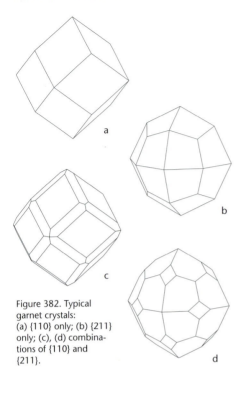

Figure 382. Typical garnet crystals: (a) {110} only; (b) {211} only; (c), (d) combinations of {110} and {211}.

Crystallography. Cubic $4/m\overline{3}2/m$; crystals common, most often in {110}, {211}, or combinations thereof; also fine to coarse granular.

Physical properties. No cleavage, conchoidal fracture. Hardness about 7; density 3.5–4.3, according to the composition. Most garnets are red to brown; more specifically, pyrope is deep red to almost black; almandine red to brown; spessartine orange, red or brown; grossular white, yellow, pink, green, or brown; andradite yellow, green, brown, or black; uvarovite emerald-green; vitreous lustre, sometimes resinous; transparent to translucent.

Chemical properties, etc. The general formula for the garnets is $A_3B_2(SiO_4)_3$, in which A includes relatively large divalent ions such as Mg^{2+}, Fe^{2+}, Mn^{2+}, or Ca^{2+}, and B relatively small trivalent ions such as Al^{3+}, Fe^{3+}, or Cr^{3+}. A and B refer to positions in the crystal struc-

Figure 384. Spessartine from Nathrop, Colorado, USA. Field of view: 14×20 mm.

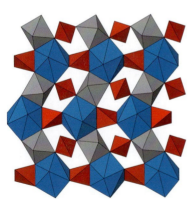

Figure 383. A slice of the crystal structure of garnet viewed along the a axis. Mg^{2+}, Fe^{2+}, Mn^{2+}, or Ca^{2+} (A atoms in the formula) are surrounded by eight oxygen atoms at the corners of a distorted cube (blue); Al^{3+}, Fe^{3+}, or Cr^{3+} (B atoms in the formula) are at the centre of an octahedron (grey); Si^{4+} ions are located in isolated tetrahedra (red).

Figure 385. Almandine from Broken Hill, New South Wales, Australia. Field of view: 40 × 61 mm.

ture in which A is surrounded by eight O atoms at the corners of a polyhedron resembling a distorted cube, and B is surrounded by six O atoms in octahedral positions. Si is in isolated tetrahedra and is normally not replaced by other elements. The group includes the following end-members:

Pyrope, $Mg_3Al_2(SiO_4)_3$; *almandine*, $Fe_3Al_2(SiO_4)_3$; *spessartine*, $Mn_3Al_2(SiO_4)_3$; *grossular*, $Ca_3Al_2(SiO_4)_3$; *andradite*, $Ca_3Fe_2(SiO_4)_3$, *uvarovite*, $Ca_3Cr_2(SiO_4)_3$, and *schorlomite*, $Ca_3(Ti,Fe)_2[(Si,Fe)O_4]_3$.

The garnets are often divided into two series, *pyralspite* (pyrope, almandine, spessartine) and *ugrandite* (uvarovite, grossular, andradite). Considerable solid solution exists within each series, but the miscibility between the two series is very limited.

Names and varieties. *Pyrope* is derived from the Greek word for fire; *almandine* from Alabanda in Turkey; *spessartine* from Spessart in Germany; *grossular* from *Ribes grossularia*, the gooseberry, owing to its green colour; *andradite* is named after the Portuguese mineralogist d'Andrada; *uvarovite* after the Russian Count Uvarov; *schorlomite*, is named after its similarity to schorl, a black opaque variety of tourmaline. *Demantoid* is a green gemstone variety of andradite; *melanite* is a black Ti-rich variety of andradite; and; *hydrogrossular* is a group of H_2O-containing garnets; *hessonite* is a yellowish- to reddish-brown and *tsavorite* is a green gemstone variety of grossular that contains V and Cr.

Occurrence. Pyrope occurs in ultrabasic rocks such as peridotites and in serpentinites derived from them; it is also found in high-grade metamorphic rocks rich in Mg. Alman-

Figure 386. Grossular from Sandaré, Nioro du Sahel, Mali. Subject: 29 × 32 mm.

dine is common in metamorphic rocks such as gneisses and schists and also in sediments. Spessartine is found in granite pegmatites and also in metamorphic Mn-rich rocks. Grossular and andradite are typically found in contact-metamorphosed or regionally metamorphosed impure limestones, commonly in association with other skarn minerals and sulphides. Schorlomite is found in alkaline rocks. Uvarovite is not as common as other garnets; it occurs in Cr-bearing serpentinites.

Use. Garnets are useful as abrasives because of their relatively high hardness in combination with a lack of cleavage. Transparent and flawless garnets are popular as gemstones; the green demantoid is the most appreciated.

Diagnostic features. Garnets are identified by their crystal habit, mode of occurrence, and, partly, by their colour.

Other nesosilicates

Zircon, $ZrSiO_4$

Crystallography. Tetragonal $4/m2/m2/m$; crystals common, usually as simple combinations of prism and bipyramid, e.g. {100} and {101}; more face-rich crystals are also found.

Physical properties. Cleavage poor on {100}. Hardness 7½; density about 4.7, decreasing with increasing metamictization. Natural zircon is normally brownish-yellow to brownish-red, whereas gemstones (often heat treated) can be colourless, yellow, or blue; adamantine lustre and very high birefringence; transparent to translucent.

Figure 387. Grossular from Jeffrey Mine, Asbestos, Quebec, Canada. Field of view: 26 × 39 mm.

Figure 388. Andradite var. melanite from the Gardiner complex, Greenland. Field of view: 54 × 87 mm.

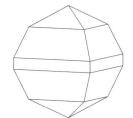

Figure 389. Zircon as in Figure 390: {110}, {112}, and {221}.

Chemical properties, etc. Zircon normally contains small amounts of Hf, sometimes also U or Th, which is the cause of the metamict state. Zr is surrounded by eight O atoms in a tetragonal polyhedron, alternating with isolated [SiO$_4$] tetrahedra along the *c* axis like pearls on a string. The entire crystal structure consists of such 'strings' displaced in relation to each other in a tetragonal arrangement.

Names and varieties. *Hyacinth* is a name sometimes used for yellowish- or brownish-red zircon gemstones.

Occurrence. Zircon is a common accessory mineral in igneous rocks, particularly in granites, syenites, and nepheline-syenites and the associated pegmatites. Particularly large crystals are found near Betroka in Madagascar and at Renfrew, Ontario, Canada. Because of its great hardness and resistance to chemical attack, zircon is also found in many sediments in accumulations with other relatively heavy minerals. Such deposits are found in, e.g. Sri Lanka, Australia, and the Urals, Russia.

Use. Zircon is the principal source of zirconium. Pure Zr is used in the construction of nuclear reactors; ZrO$_2$, having an extremely high melting point, is used, e.g. for crucibles for melting platinum and for ceramics. Clear zircons are used as gemstones; many zircons have been heat-treated to change their colour. The presence of radioactive elements in zircon makes it useful for determining the radiometric ages of rocks.

Diagnostic features. Crystal habit, hardness, density, lustre, and colour.

Thorite, (Th,U)SiO$_4$, and coffinite, U[SiO$_4$,(OH)$_4$]

Tetragonal with a zircon structure. These minerals are generally dark-coloured to black, and have a high density and high lustre. They are radioactive and are normally metamict. Their mode of occurrence is almost the same as for zircon.

Figure 390. Zircon from Seiland, Finnmark, Norway. Subject: 34 × 40 mm.

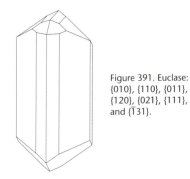

Figure 391. Euclase: {010}, {110}, {011}, {120}, {021}, {111}, and {$\bar{1}$31}.

Euclase, BeAlSiO$_4$(OH)

Monoclinic and found in well-developed crystals, usually prismatic parallel to the *c* axis. Hardness 7½; density 3.0. Cleavage perfect on {010}. Colourless to light green or blue; vitreous lustre. Euclase is found in granitic pegmatites and is occasionally used as a gemstone.

The Al$_2$SiO$_5$ group

The Al$_2$SiO$_5$ group includes sillimanite, andalusite, and kyanite, which are characteristic minerals in Al-rich metamorphic rocks. They serve as valuable tools in the interpretation of geological events, because they are indicators of the pressure and temperature conditions during metamorphic processes. Figure 392 shows the stability conditions of the three minerals. The minerals are polymorphs, i.e. they have the same chemical formula, Al$_2$SiO$_5$, but with different crystal structures. *Mullite*

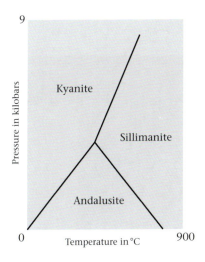

Figure 392. Stability diagram for the three Al$_2$SiO$_5$ minerals. The pressure and temperature conditions for the point where the three phases meet correspond to a depth of 15–20 km in the earth's crust

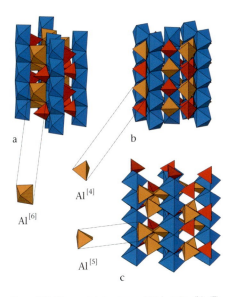

Figure 393. The crystal structures of (a) kyanite, (b) sillimanite, and (c) andalusite, all with the formula Al$_2$SiO$_5$. All three minerals have Si in isolated tetrahedra (red) and half the Al in chains of octahedra (blue) along the *c* axes. The coordination number, i.e. the number of O atoms as nearest neighbours, for the remaining Al atoms (yellow) is different for each of the three minerals; in kyanite it is six, in sillimanite four, and in andalusite five. Kyanite, formed under high pressure, has the densest structure, which is also reflected in a higher density.

263

Figure 394. Sillimanite from Purnamoota Road, New South Wales, Australia. Subject: 51 × 63 mm.

is a rare mineral in nature but a common phase produced by heating sillimanite, andalusite, or kyanite. It is found in porcelain and ceramics.

Sillimanite, Al_2SiO_5

Crystallography. Orthorhombic $2/m2/-m2/m$, good crystals rare, {110} usually the predominant form; mostly fibrous, columnar.

Physical properties. Cleavage perfect on {010}. Hardness 6–7; density 3.2. Colour is greyish, yellowish, light brown, or green; vitreous lustre, silky when fibrous; transparent to translucent.

Chemical properties, etc. The crystal structure of sillimanite consists of chains of $[AlO_6]$ octahedra linked by alternating $[SiO_4]$ tetrahedra and $[AlO_4]$ tetrahedra.

Names and varieties. *Fibrolite* is an obsolete name for sillimanite.

Occurrence. Sillimanite primarily occurs in Al-rich high-grade regional metamorphic rocks such as gneisses and micaschists; it is also found in contact-metamorphic rocks.

Diagnostic features. Sillimanite is best known by its fibrous or columnar habit. It differs from other fibrous silicates like wollastonite or tremolite by having only one direction of cleavage.

Andalusite, Al_2SiO_5

Crystallography. Orthorhombic $2/m2/-m2/m$. Crystals normally simple prisms with a nearly square cross-section; surfaces often slightly altered to a mica-like product.

Physical properties. Cleavage good on {110}. Hardness 7½; density 3.2. Colour is usually grey or dirty green, brown, or red; vitreous lustre; translucent, rarely transparent; strong pleochroism in transparent crystals.

Chemical properties, etc. The crystal structure of andalusite consists of chains of $[AlO_6]$ octahedra linked by alternating $[SiO_4]$ tetra-

Figure 395. Andalusite from the former Lizenseralp (Tirol), Italy. Field of view: 42 × 47 mm.

Figure 396. Andalusite var. chiastolite from Lancaster, Massachusetts, USA. Field of view: 16 × 24 mm.

hedra and [AlO$_5$] polyhedra. In small amounts Mn^{3+} can replace Al.

Names and varieties. *Chiastolite* is an andalusite with dark carbonaceous inclusions seen as cruciform patterns on a prismatic cross-section.

Occurrence. Andalusite is a characteristic mineral in Al-rich contact-metamorphic rocks, e.g. shales. It is also found in regionally metamorphosed rocks. In addition to Andalusia in Spain, which gives its name to the mineral, valuable deposits are found in California, USA. Andalusite of gemstone quality is known from Santa Teresa in Espirito Santo, Brazil.

Use. Andalusite has been used in the production of porcelain for sparking-plugs and similar materials. Transparent varieties are used as gemstones.

Diagnostic features. Andalusite is identified its by mode of occurrence, crystal habit, hardness, and cruciform inclusions.

Kyanite, Al$_2$SiO$_5$

Crystallography. Triclinic $\bar{1}$; crystals common, usually elongated after the *c* axis and tabular parallel to {100} with subordinate {010}; {100} faces sometimes horizontally striated; twinning, e.g. on {100}, occurs; also in bladed aggregates.

Physical properties. Cleavage perfect on {100}. Hardness varies with crystallographic direction: on {100} about 4½ parallel to the *c* axis, but normal to this about 6½; density 3.6. Colour is usually bluish, often with increasing intensity towards the central parts of a crystal, also whitish, greyish, or greenish; vitreous to pearly lustre; transparent to translucent.

Chemical properties, etc. The crystal structure of kyanite consists of chains of [AlO$_6$] octahedra linked by [SiO$_4$] tetrahedra and [AlO$_6$] octahedra.

Names and varieties. Kyanite was formerly

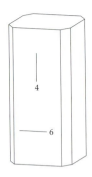

Figure 397. Kyanite: {100}, {010}, {001}, {110}, and {1$\bar{1}$0}. The hardness of kyanite varies with the crystallographic direction: along the c axis it is 4-4½; perpendicular to the c axis it is 6–7.

called *disthene* with reference to its variation in hardness with crystallographic direction.

Occurrence. Kyanite is found in Al-rich regional metamorphic rocks such as mica-schists and gneisses, usually in association with staurolite and garnet. It also occurs in eclogites and comparable rocks formed under extreme high pressure. Fine kyanite crystals are found at St. Gotthard in Switzerland, in Yancy county in North Carolina, USA, and at several Brazilian localities.

Use. Like andalusite, kyanite has been used in the production of porcelain for sparking-plugs and similar materials.

Diagnostic features. The hardness properties, blue colour with variable intensity, crystal habit, and cleavage make kyanite easy to identify.

Figure 398. Kyanite in quartz from Brazil. Field of view: 80 × 83 mm.

Staurolite,
$(Fe,Mg)_4Al_{17}(Si,Al)_8O_{45}(OH)_3$

Crystallography. Monoclinic $2/m$, pseudo-orthorhombic; commonly in crystals, usually in combinations of {110} and {001}, {010} and {101}; twinning common according to two twin laws, {031} and {231}, both forming cruciform twins.

Physical properties. Cleavage distinct on {010}. Hardness 7; density 3.7. Colour is light brown, reddish-brown, or brownish-black; vitreous lustre, sometimes resinous; translucent to nearly opaque.

Chemical properties, etc. The chemical content varies somewhat, e.g. Fe^{3+} can partly replace Al. The crystal structure has features in common with that of kyanite, since staurolite

Figure 400. Staurolite, twinned on {031}, from St. Gotthard, Switzerland. Subject: 29×28 mm.

Figure 401. Staurolite in paragonite-schist from St. Gotthard, Switzerland. Field of view: 26×37 mm.

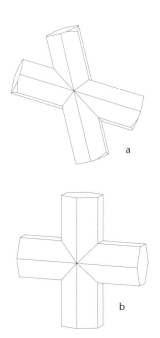

Figure 399. Twins of staurolite: cruciform twins, (a) twinned on {231} with 60° between the individuals; (b) twinned on {031} with 90° between the individuals.

consists of layers of kyanite alternating with a different type of layer. These layers are parallel to {010}, which explains how crystals of kyanite and staurolite can grow together in a crystallographic orientation in which they share a {010} face and have their c axes parallel.

Names and varieties. The name is from the Greek word *stauros*, 'cross', with reference to the characteristic twins.

Occurrence. Staurolite is found in Al-rich regionally metamorphosed rocks, such as mica-schists and gneisses, usually in association with kyanite, garnet, or sillimanite. Fannin County in Georgia, USA, Monte Campione in Switzerland, and Rubelita in Minas Gerais, Brazil, are among the well-known localities.

Diagnostic features. Colour, crystal habit, and the special mode of twinning are characteristic.

Topaz, $Al_2SiO_4(F,OH)_2$

Crystallography. Orthorhombic $2/m2/m2/m$; crystals common, usually short to long prismatic along the c axis; crystals terminated by bipyramids, other prisms, or pinacoid; prism faces often vertically striated; also fine to coarse granular.

Physical properties. Cleavage perfect on {001}. Hardness 8; density 3.5. Colourless or in yellow, brown, blue, or shades of green, rarely pink; vitreous lustre; transparent or translucent.

Chemical properties, etc. F usually predominates over OH; the crystal structure consists of chains of $[AlO_4F_2]$ octahedra parallel to the c axis, linked by isolated $[SiO_4]$ tetrahedra. This structure is reflected in the common prismatic crystal habit. Cleavage can take

Figure 402. Topaz from Mursinka, Urals, Russia. Field of view: 40×60 mm.

place by breaking only Al–O and Al–F bonds without breaking Si–O bonds. The structure is a relatively dense packing of atoms, which explains the relatively high density.

Occurrence. Topaz is a characteristic mineral in pneumatolytic and pegmatitic veins related to Si-rich igneous rocks such as granites and rhyolites. Associated minerals are typically albite, tourmaline, cassiterite, apatite, fluorite, beryl, and micas.

At Ouro Preto and other localities in Minas Gerais, Brazil, very large crystals up to more than 200 kg have been found. In Russia fine specimens are found in several places in the Nerchinsk district in Siberia and at the classic occurrence at Mursinka in the Urals, Russia. In the USA topaz is found, e.g. at Pikes Peak in Colorado and the Thomas Range in Utah. Topaz is known also from Sri Lanka and Pakistan and the classic Schneckenstein occurrence in Saxony, Germany.

Use. Topaz is an appreciated gemstone. Fine cut stones can be expensive, which has led to the sale of similar yellow or brown minerals, especially quartz, under fanciful names with topaz as a root name.

Diagnostic features. Crystal habit, great hardness, high density, and mode of occurrence. Danburite is similar to topaz in crystal habit and physical properties.

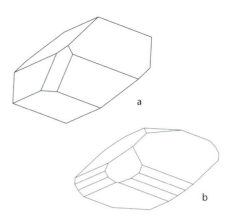

Figure 404. Titanite: (a) {001}, {011}, {$\bar{1}$01}, and {$\bar{1}$23}; (b) {001}, {110}, {011}, {$\bar{1}$01}, {$\bar{1}$02}, {121}, and {$\bar{1}$23}.

Titanite, $CaTiO(SiO_4)$

Crystallography. Monoclinic 2/m; crystals typically flattened parallel to {001}, wedge-shaped; common forms are {100}, {001}, {110}, and {111}; twinning on {100} giving contact or penetration twins; also granular, lamellar.

Physical properties. Cleavage distinct on {110}. Hardness 5–5½; density 3.4–3.6. Colour is usually brown, also yellow, green, or grey to black; weak adamantine lustre; transparent or translucent.

Chemical properties, etc. The chemical composition varies slightly. Ca can be partially replaced by Na or rare-earth elements, Ti by, e.g. Al, Fe, or Nb, and Si by Al. F can replace O in small amounts. The crystal structure consists of chains of $[TiO_6]$ octahedra linked by isolated $[SiO_4]$ tetrahedra; this linkage results in a framework with relatively large cavities, which Ca occupies.

Names and varieties. Titanite is named after its content of titanium. *Sphene* is an obsolete name for titanite.

Occurrence. In small amounts titanite is widespread in igneous rocks from granites to nepheline-syenites; it is also found in some

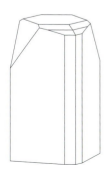

Figure 403. Topaz as in Figure 402: {001}, {110}, {120}, {021}, {113}, and {114}.

Figure 405. Titanite in natrolite from the Gardiner complex, Greenland. Field of view: 24 × 30 mm.

metamorphic rocks, such as chlorite-schists or crystalline limestones. There is a large deposit of titanite in the Lovozero and Khibina massifs, Kola Peninsula, Russia, where it is found together with apatite and nepheline. Fine specimens are known from St. Gotthard and other localities in the Alps as well as from Tilly Foster mine in New York State and Bridgewater in Pennsylvania, USA.

Use. Titanite, where present in large amounts, is mined as a source of Ti.

Diagnostic features. Relatively high lustre, flattened crystal habit; colour varies, but is usually diagnostic.

Chloritoid, $(Fe,Mg,Mn)Al_2SiO_5(OH)_2$

Monoclinic or triclinic. Sometimes in simple tabular crystals, commonly twinned; occurs mostly in scaly or foliated forms. Cleavage good on {001}; hardness 6½; density about

Figure 406. Chloritoid in quartz from Prägraten, Tirol, Austria. Field of view: 36 × 60 mm.

Figure 407. Datolite from Asdal, Arendal, Norway. Field of view: 40 × 60 mm.

3.6. Colour is usually greenish-grey to black; vitreous lustre, pearly on cleavage surfaces. Chloritoid is found in low- to medium-grade metamorphic rocks such as schists in association with chlorite, muscovite, and garnet. It resembles chlorite and owes its name to this resemblance, but can be distinguished from chlorite by its greater hardness and by having very brittle cleavage folia.

Datolite, $CaBSiO_4(OH)$

Monoclinic and found as short prismatic crystals, commonly with many faces, or in granular porcelain-like masses. No distinct cleavage; hardness about 5; density 3.0. Colourless, white, or pale greenish; vitreous lustre, somewhat pearly. Datolite, a member of the gadolinite group, occurs in cavities in basaltic lavas, usually in association with zeolites and calcite. It is best identified by its mode of occurrence, crystal habit, or porcelain-like appearance. *Homilite*, $Ca_2(Fe,Mg)B_2Si_2O_{10}$, and *hingganite*, $(Yb,Y)_2Be_2Si_2O_8(OH)_2$, are closely related minerals that occur in nepheline-syenites.

Figure 408. Kornerupine in anorthosite from Qeqertarsuatsiaat (Fiskenæsset), Greenland. Field of view: 50 × 75 mm

Figure 409. Dumortierite from Madagascar. Subject: 42 × 64 mm.

Gadolinite, $Be_2FeY_2Si_2O_{10}$

Monoclinic; rarely as crystals, mostly in dense masses with conchoidal fracture. Hardness 6½; density 4–4.7. Colour is brown to black, streak greenish-grey; greasy lustre, and nearly opaque. Y is often partially replaced by rare-earth elements or Th, and the Th content normally makes gadolinite metamict. Gadolinite occurs in granitic pegmatites, e.g. at Ytterby, Sweden, and Iveland, Norway, where large crystals are found. Gadolinite is known by its metamict appearance and its mode of occurrence, including the presence of fissures in the neighbouring minerals because of the expansion caused by the metamictization.

Kornerupine, $Mg_4Al_6(Si,Al,B)_5O_{21}(OH)$

Orthorhombic and found as long prismatic crystals in columnar or radiated aggregates. Cleavage on {110}; hardness 6½; density 3.3. Colourless or with pale yellow, brown, or green colours; transparent or translucent. Named after the Danish geologist Andreas Kornerup. *Prismatine* is a closely related mineral. Kornerupine occurs in Mg–Al-rich metamorphic rocks, e.g. associated with sapphirine and cordierite at Qeqertarsuatsiaat (Fiskenæsset), Greenland. It is also known from, e.g. Madagascar and Sri Lanka, where it is found as transparent crystals suitable as gemstones.

Dumortierite, $(Al,Mg,Fe)_{27}B_4Si_{12}O_{69}(OH)_3$

Orthorhombic; found in prismatic crystals, but mostly in dense fibrous masses. Hardness 7; density 3.3. Colour is dark blue to violet, greyish-blue, brownish, or reddish; silky lustre; translucent. Dumortierite is primarily found in granitic pegmatites. Its colour and fibrous appearance are characteristic.

Figure 410. Cuprosklodowskite from Shaba, Congo. Field of view: 24 × 34 mm.

Uranophane, $Ca(UO_2)_2(SiO_3OH)_2 \cdot 5H_2O$

Monoclinic; found as small needle-shaped crystals, commonly in radiating spheres; also as felted aggregates. Cleavage good on {100}; hardness 2½; density 3.9. Colour is yellow to orange-yellow; vitreous lustre, somewhat pearly; translucent. Ura·nophane is a secondary mineral occurring in pegmatites as an alteration product of uraninite. Closely related are *kasolite*, $Pb(UO_2)SiO_4 \cdot H_2O$, with Pb replacing Ca, and *cuprosklodowskite*, $Cu(UO_2)_2(SiO_3OH)_2 \cdot 6H_2O$ with Cu replacing Ca; the latter is typically found as green needle-shaped crystals. These minerals are of economic importance as ores of U.

273

Figure 411. Epidote from Castrovirreyna, Huancavelica, Peru. Subject: 55 × 65 mm.

Sorosilicates

In sorosilicates the $[SiO_4]$ tetrahedra are linked in pairs. A few of the minerals in this group have $[SiO_4]$ tetrahedra both in pairs and isolated. In consequence, the silicate part of the formula can be more complex than Si_2O_7, which applies to the simple cases.

Thortveitite, $(Sc,Y)_2Si_2O_7$

Monoclinic, found as prismatic crystals, sometimes rather large. It is a rare, dark green to black mineral with hardness 6½ that is found in granitic pegmatites, e.g. at Iveland, Norway, and Crystal Mountain mine in Montana, USA.

Melilite, $(Ca,Na)_2(Al,Mg)(Si,Al)_2O_7$

Crystallography. Tetragonal $\bar{4}2m$; crystals small, usually simple with only {110} and {001}; mostly granular.
Physical properties. Cleavage distinct on {001}, indistinct on {110}. Hardness 5½;

density about 3.0. Colour is yellowish, brownish, greyish, or colourless; vitreous lustre, somewhat greasy on fresh surfaces; transparent to translucent.

Chemical properties, etc. Melilite is the common member of a solid-solution series between *åkermanite*, $Ca_2MgSi_2O_7$, and *gehlenite*, $Ca_2Al(Si,Al)_2O_7$. In melilite some Ca is usually replaced by Na, whereas minor Fe can replace Mg or Al. In the crystal structure of melilite the paired $[SiO_4]$ tetrahedra are arranged in (001) layers and linked by $[(Al,Mg)O_4]$ tetrahedra; the larger cations Ca and Na are accommodated between these layers, which explains the cleavage.

Occurrence. Melilite is a rock-forming mineral found in igneous rocks that are very poor in Si and rich in Ca, in which it replaces feldspar. It is found, e.g. in the Gardiner complex, Greenland, and at Iron Hill, Colorado, USA. Melilite also occurs in contact-metamorphosed limestones.

Figure 412. Melilite from the Gardiner complex, Greenland. Subject: 83 × 98 mm.

Use. Natural melilite is not mined but melilite-like phases are important constituents in cement and furnace slags.

Diagnostic features. Melilite in granular masses is difficult to identify. An old weathered surface is often yellowish-coated, which provides a good clue.

Leucophanite, $NaCaBeSi_2O_6F$ and meliphanite, $Na(Na,Ca)BeSi_2O_6F$

Leucophanite and meliphanite are Be-containing minerals related to melilite; they usually occur as yellow or greenish-yellow tabular crystals and are known from, e.g. the nepheline-syenite pegmatites at Mont Saint-Hilaire, Quebec, Canada.

Figure 413. Bertrandite on quartz from Kara-Oba, Kazakhstan. Field of view: 50×73 mm.

Bertrandite, $Be_4Si_2O_7(OH)_2$

Orthorhombic and found in small tabular crystals notable for their hemihedral symmetry. Cleavage perfect on {001}, good on {110}, and {101}; hardness 6½; density 2.6. Colourless or yellowish; vitreous lustre, pearly on cleavage surfaces; transparent to translucent. Bertrandite occurs in granitic pegmatites, commonly as a hydrothermal alteration product of beryl. Bertrandite is not easy to identify, but the association with beryl is usually characteristic. Occasionally bertrandite is found in distinctive heart-shaped twins.

Hemimorphite, $Zn_4Si_2O_7(OH)_2 \cdot H_2O$

Crystallography. Orthorhombic $mm2$; crystals common, commonly tabular parallel to {010} and with a distinct hemihedral development of the terminations of the c axis. Twinning on {001} occurs; crystals are commonly in fan-shaped groups almost parallel to {010}; also stalactitic, botryoidal, or as coatings.

Physical properties. Cleavage perfect on {110}. Hardness 5; density about 3.4. Colour is white, also bluish, greenish, yellowish, or brownish; vitreous lustre; transparent to translucent. Strongly pyroelectric and piezoelectric.

Chemical properties, etc. In the crystal structure of hemimorphite the paired [SiO_4] tetrahedra are linked to other tetrahedra with Zn. All the [SiO_4] tetrahedra have their bases parallel to {001} and their apexes pointing in the same direction. This polar structure is responsible for the hemihedral crystal habit and the electrical properties.

Names and varieties. *Calamine* is an obsolete name for hemimorphite.

Occurrence. Hemimorphite is found in the oxidized zones of Zn deposits, often in association with other Zn minerals such as smithsonite and sphalerite. Exquisite crystals of hemimorphite are found in Santa Eulalia, Chihuahua, and other places in Mexico.

Use. Hemimorphite is an inferior zinc ore.

Figure 414. Hemimorphite: {100}, {010}, {001}, {110}, {011}, {101}, {301}, {031}, {121}, and {12$\bar{1}$}.

Axinite, $Ca_2(Mn,Fe,Mg)Al_2BSi_4O_{15}(OH)$

Crystallography. Triclinic $\bar{1}$; crystals common, commonly axe-head-shaped; also massive, granular.

Physical properties. Cleavage good in one direction. Hardness 6½–7; density 3.3. Colour is usually brown to violet, also yellowish, greenish, or greyish; vitreous lustre; transparent or translucent.

Diagnostic features. The hemimorphic habit (which gives its name to the mineral), the fan-shaped crystal groups, and the electrical properties.

Figure 416. Axinite: {100}, {110}, {1$\bar{1}$0}, {$\bar{2}$01}, {401}, and {53$\bar{1}$}.

Figure 415. Hemimorphite from Mapimi, Durango, Mexico. Subject: 54 × 75 mm.

Figure 417. A small section from the crystal structure of axinite. Isolated [BO$_4$] tetrahedra (yellow) link pairs of [SiO$_4$] tetrahedra (red) together to build small clusters.

Chemical properties, etc. The content of Ca, Mn, Mg, and Fe varies, and axinite is in fact a small group of minerals including *ferro-axinite* and *manganaxinite*. In axinite the paired [SiO$_4$] tetrahedra are linked by isolated [BO$_4$] tetrahedra to form small clusters. The clusters are arranged in layers alternating with layers of octahedra with Ca and other cations.

Occurrence. Axinite occurs in veins and cavities in granites, especially in the contact zones with country rocks, typically limestones. Fine crystals are known from, e.g. Bourg d'Oisans, France, and Coarse Gold, California, USA.

Diagnostic features. Crystal habit and colour.

Lawsonite, $CaAl_2Si_2O_7(OH)_2 \cdot H_2O$

Orthorhombic; found in prismatic or tabular crystals, mostly granular. Cleavage perfect on {001} and {100}, indistinct on {110}; hardness 6; density 3.1. Colour is white, pale blue, or greyish-blue; vitreous lustre. Lawsonite has nearly the same chemical composition as anorthite, but a more dense crystal structure gives it a higher density. It is a typical mineral in glaucophane-schists and similar low-grade metamorphic rocks. Its chief characteristics are cleavage, density, and especially mode of occurrence.

Figure 418. Axinite from Puiva, Urals, Russia. Field of view: 25 × 37 mm.

Ilvaite, $CaFe^{3+}(Fe^{2+})_2O(Si_2O_7)(OH)$

Monoclinic and found as long prismatic crystals, prism faces are commonly vertically striated; also in massive or radiated aggregates. Cleavage distinct on {010} and {001}; hardness 5½–6, density about 4.0. Colour is black or brownish-black with a nearly black streak; vitreous lustre, sometimes submetallic. Ilvaite is a typical mineral found in contact-metamorphosed rocks and is known, e.g. in dolomite on Elba, Italy. Fine ilvaite crystals are also known from Dalnegorsk, Russia. Radiating aggregates of ilvaite can resemble tourmaline or black amphiboles.

Låvenite, $Na_2MnZr(Si_2O_7)(O,F)_2$

Monoclinic and found mostly as long prismatic crystals with predominant {110}; twinning on {100} common. Cleavage good on {100}; hardness 6; density 3.5. Colour is yellow to brown; vitreous lustre. Låvenite belongs to the *låvenite–wöhlerite group*, a small group of minerals whose members are structurally related and have rather uniform properties. The chemical composition varies, e.g. in Na being partly or entirely replaced by Ca, Mn, or Fe, and Zr by Nb or Ti. *Wöhlerite* is usually honey-yellow and is found in tabular crystals; *mosandrite* has a colour like låvenite and occurs mostly in granular aggregates or poorly developed crystals; the more greyish *rosenbuschite*, a related mineral, is known by its needle-shaped crystals in small fan-shaped aggregates. These minerals are found in

Figure 419. Ilvaite from Dalnegorsk, Russia. Subject: 41 × 50 mm.

Figure 420. Mosandrite in hornblende from Langesundsfjorden, Norway. Field of view: 66 × 126 mm.

Figure 421. Lamprophyllite (yellow) and amphibole (black) in natrolite from the Gardiner complex, Greenland. Subject: 58 × 84 mm.

nepheline-syenite pegmatites, e.g. at Lange-sundsfjorden, Norway, and Mont Saint-Hilaire, Quebec, Canada.

Lamprophyllite, $Sr_2Na_3Ti_3(Si_2O_7)_2(OH)_4$

Monoclinic and usually found as tabular or ruler-shaped crystals, rarely needle-shaped. Cleavage perfect on {100}; hardness 2½; density 3.3. Colour is golden-yellow to brownish; strong vitreous lustre, occasionally submetallic. The chemical composition can vary, e.g. by Sr being partially replaced by Ba. Lamprophyllite occurs in nepheline-syenites and associated pegmatites, e.g. in the Khibina and Lovozero massifs in the Kola Peninsula, Russia. Lamprophyllite is usually readily

identified by its mica-like appearance and its mode of occurrence, but can nevertheless be confused with astrophyllite.

Murmanite, $Na_3(Ti,Nb)_4O_4(Si_2O_7)_2 \cdot 4H_2O$

Triclinic; mostly as tabular crystals parallel to {100}. Cleavage perfect on {100}; cleavage blades brittle. Hardness 2½, density 2.8. Colour is violet to bronze-like; vitreous to submetallic lustre. *Epistolite*, white, and *vuonnemite*, yellowish, are closely related minerals. All three minerals are found in nepheline-syenites, e.g. in the Ilímaussaq complex, Greenland, and corresponding complexes in the Kola Peninsula, Russia.

Figure 422. Murmanite in a lujavrite (a type of feldspathoidal syenite) from Kvanefjeld, the Ilímaussaq complex, Greenland. Field of view: 60 × 69 mm.

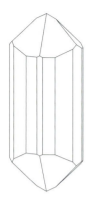

Figure 423. Danburite: {100}, {010}, {110}, {011}, {210}, {410}, and {412}.

Danburite, $CaB_2Si_2O_8$

Orthorhombic; mostly found in topaz-like prismatic crystals. Conchoidal or uneven fracture; hardness 7–7½, density about 3. Colourless or light yellowish, greasy vitreous lustre. Danburite is found in contact-metamorphic rocks and in ore deposits formed at relatively high temperatures. It resembles topaz and is occasionally seen as a gemstone. Fine crystals are known, e.g. from Charcas, San Luis Potosi, Mexico.

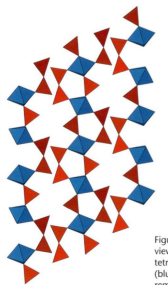

The epidote group

The epidote group includes *epidote, clinozoisite, piemontite, allanite, and zoisite.* These minerals are mainly found in metamorphic rocks, but allanite occurs mostly in granites and granitic pegmatites. They have basically the same monoclinic crystal structure, except for zoisite, which is orthorhombic. The crystal structure of epidote consists of both paired and isolated $[SiO_4]$ tetrahedra. They are linked by $[AlO_6]$ octahedra that form infinite chains along the *b* axis. This framework offers space for a large cation, Ca, surrounded by eight O atoms and a minor one in octahedral position, i.e. surrounded by six O atoms; this latter position is in epidote occupied by Fe^{3+}, in zoisite and clinozoisite by Al^{3+}, and in piemontite by Mn^{3+}.

Epidote, $Ca_2FeAl_2(Si_2O_7)(SiO_4)(O,OH)_2$

Crystallography. Monoclinic $2/m$; crystals commonly elongated parallel to the *b* axis and dominated by {100}, {001}, and other forms parallel to this axis. Often these faces are also striated parallel to the *b* axis; twinning on {100} occurs; also granular, rarely fibrous.

Physical properties. Cleavage good on {001}, less good on {100}. Hardness 6–7; density about 3.4. Colour is yellowish-green, dark green, brownish-green, or nearly black; vitreous lustre; usually translucent, rarely transparent.

Chemical properties, etc. A complete solid-solution series exists between epidote, the

Figure 424. Parts of the crystal structure of epidote viewed along the *b* axis. Paired and isolated $[SiO_4]$ tetrahedra (red) are linked together by $[AlO_6]$ octahedra (blue), which form infinite chains along the *b* axis. The remaining positions in the structure are not shown.

Figure 425. Epidote: {100}, {010}, {001}, {110}, {011}, {012}, {210}, {10$\bar{1}$}, {$\bar{2}$01}, {$\bar{3}$01}, {$\bar{3}$04}, {$\bar{1}$11}, {$\bar{2}$11}, and {$\bar{2}$33}.

Fe^{3+} end-member, and clinozoisite, the Al^{3+} end-member.

Names and varieties. *Pistacite*, referring to the characteristic pistachio-green colour, is an obsolete name for epidote.

Occurrence. Epidote is an abundant mineral and is especially widespread in regionally metamorphosed rocks of low to moderate grade; in these rocks it is typically associated with Ca-rich amphiboles, albite, quartz, and chlorite. It also occurs in contact-metamorphosed limestones, particularly where these are associated with Fe ore; typical minerals in this association are garnets, vesuvianite, diopside, and calcite. Epidote is also common in thin veins and fissures in granites, and is found together with zeolites as late fillings in cavities in lavas. Splendid epidote crystals are found at Knappenwand, Untersulzbachtal, Austria, on Prince of Wales Island, Alaska, USA, and recently at several localities in Pakistan.

Use. Epidote is occasionally seen as a gemstone.

Diagnostic features. Colour, crystal habit, cleavage, and, usually, the mineral association and mode of occurrence.

Figure 426. Epidote from Prince of Wales Island, Alaska, USA. Field of view: 50 × 52 mm.

Figure 427. Clinozoisite from Alchuri, Gilgit, Pakistan. Subject: 52 × 67 mm.

Clinozoisite,
$Ca_2Al_3(Si_2O_7)(SiO_4)(O,OH)_2$

A member of the epidote group, forming a complete solid-solution series with epidote. It resembles epidote both in properties and in mode of occurrence, but is generally lighter coloured. *Zoisite* has the same formula as clinozoisite, but is orthorhombic. It is typically greyish, but is also found in a pinkish variety, *thulite*, and in a blue gemstone variety, *tanzanite*, from Tanzania.

Piemontite,
$Ca_2MnAl_2(Si_2O_7)(SiO_4)(O,OH)_2$

Crystal structure as for epidote, but with Mn^{3+} replacing Fe^{3+}. Piemontite normally occurs in radiating or granular aggregates with a reddish-brown or reddish-black colour; streak reddish. Piemontite is found in association with Mn-rich ore deposits.

Allanite,
$Ca(Ce,La)(Al,Fe,Cr,V)_3(Si_2O_7)(SiO_4)(O,OH)_2$

Crystallography. Monoclinic $2/m$; crystal habit as for epidote but generally found granular, massive; often metamict.

Physical properties. Cleavage rarely as distinct as in epidote; conchoidal or uneven fracture. Hardness 5½–6; density 3.5–4.2, according to the degree of metamictization. Colour is black or dark brown, streak dark brown; vitreous, 'pitchy' lustre, sometimes greasy or submetallic; partly translucent. Weakly radioactive.

Chemical properties, etc. Allanite has a more complicated composition than epidote. Some Ca is replaced by Ce, La, Na, or Th, and Fe^{3+} is partly replaced by Fe^{2+}, Mn^{3+}, and Mg^{2+}. The simultaneous presence of Fe^{3+} and Fe^{2+} is largely responsible for the black colour. The radioactive decay of Th is the cause of the metamictization.

Names and varieties. *Orthite* is an obsolete name for allanite.

Occurrence. In small amounts allanite is common in granitic or syenitic rocks and is sometimes found in greater masses or as crystals in the related pegmatites, often in association with epidote. Ytterby in Sweden, Arendal in Norway, and Amelia, Virginia, and Pacoima, California, USA, are among the best-known localities.

Diagnostic features. The 'pitchy' appearance, colour, and mode of occurrence are characteristic.

Pumpellyite,
$$Ca_2MgAl_2(SiO_4)(Si_2O_7)(OH)_2 \cdot H_2O$$

Monoclinic and occurs mostly as bladed or fibrous aggregates, commonly in rosettes. Cleavage distinct on {001}, less distinct on {100}; hardness 5½; density 3.2. Green, bluish-green, brown, or rarely colourless; vitreous lustre. *Pumpellyite* is the name of a group of seven minerals. The formula given above is for pumpellyite-(Mg). In other members of the group Mg is replaced by Al, Fe, or Mn. Pumpellyite typically occurs in veins and cavities in basalts and andesites in association with zeolites and, in particular, epidote-group minerals, with which it has common features.

Figure 428. Allanite from Avigait at Paamiut, Greenland. Subject: 81×95 mm.

Vesuvianite, $(Ca,Na)_{19}(Al,Mg,Fe)_{13}$-$(SiO_4)_{10}(Si_2O_7)_4(OH,F,O)_{10}$

Figure 430. Vesuvianite as in Figure 429: {100}, {001}, {110}, and {111}.

Crystallography. Tetragonal $4/m2/m2/m$; crystals common, usually prismatic in combinations of {100}, {101}, or {110}; faces commonly striated parallel to c axis; also in massive, granular, or radiating aggregates.

Physical properties. No distinct cleavage. Hardness 6½; density about 3.4. Colour is normally green or brown, rarely yellow or blue; vitreous lustre, somewhat resinous; transparent to translucent.

Chemical properties, etc. The composition varies, e.g. by Ca being partially replaced by Na, and with Al, Mg, and Fe occurring in various ratios. Some similarities exist between vesuvianite and garnets with respect to the formula as well as crystal structure.

Names and varieties. *Idocrase* is an old synonym for vesuvianite. *Wiluite* is a closely related mineral from the Wilui River in Yakutia, Russia, which was previously thought to be identical with vesuvianite.

Occurrence. Vesuvianite is a typical mineral in contact-metamorphosed impure limestones, where it occurs in association with calcite, Ca-rich garnets, diopside, wollastonite, etc. It is found, e.g. at Lowell, Vermont, and Sanford, Maine, USA, and at Vesuvius in Italy, from which it got its name.

Use. Vesuvianite is rarely seen as a gemstone; it is sometimes found in jade-like forms suitable for carving.

Diagnostic features. Well-developed crystals are easy to identify by habit; vesuvianite can otherwise be mistaken for garnet.

Figure 429. Vesuvianite from Wilui, Siberia, Russia. Field of view: 39×63 mm.

Cyclosilicates

In cyclosilicates the [SiO_4] tetrahedra are linked together to form rings. The Si/O ratio in the silicate part of the formula is 1:3, e.g. Si_3O_9 or Si_6O_{18}. The symmetry of the rings is normally reflected in the crystal symmetry, as in the case of beryl, which has [Si_6O_{18}] rings and is hexagonal.

Figure 431. Beryl var. aquamarine from Nagar, Hunza Valley, Pakistan. Subject: 60×72 mm.

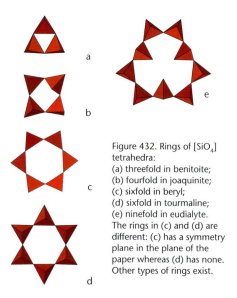

Figure 432. Rings of [SiO$_4$] tetrahedra:
(a) threefold in benitoite;
(b) fourfold in joaquinite;
(c) sixfold in beryl;
(d) sixfold in tourmaline;
(e) ninefold in eudialyte.
The rings in (c) and (d) are different: (c) has a symmetry plane in the plane of the paper whereas (d) has none. Other types of rings exist.

Figure 433. Benitoite on natrolite from Dallas Gem Mine, San Benito County, California, USA. Field of view: 28 × 42 mm.

Benitoite, BaTiSi$_3$O$_9$

Hexagonal; primarily known as well-developed crystals dominated by a trigonal bipyramid. No distinct cleavage; hardness 6½, density 3.7. Colour is light blue to dark blue; vitreous lustre; crystals often dim. Benitoite is a rare mineral that is primarily known from San Benito in California, USA, where it occurs in dykes in glaucophane-schists in association with natrolite and neptunite.

Catapleiite, Na$_2$ZrSi$_3$O$_9$·2H$_2$O

Pseudo-hexagonal; mostly found as tabular crystals dominated by {001}, commonly in rosettes. Hardness 5–6; density 2.8. Colourless to honey-yellow or brown, rarely bluish or greenish; vitreous lustre. Catapleiite occurs in nepheline-syenite pegmatites, e.g. at Langesundsfjorden, Norway, at Narssârssuk, Greenland, and Mont Saint-Hilaire, Quebec, Canada.

Joaquinite,
$NaBa_2FeTi_2Ce_2(SiO_3)_8O_2(OH) \cdot H_2O$

Monoclinic and forms tiny cube-like honey-coloured crystals. Hardness about 5; density 4.0. It occurs, e.g. with benitoite in San Benito, California, and in the Ilímaussaq complex, Greenland.

Beryl, $Be_3Al_2Si_6O_{18}$

Crystallography. Hexagonal $6/m2/m2/m$; crystals common, usually simple prismatic with $\{10\bar{1}0\}$ and $\{0001\}$ only, less commonly with several crystal forms; prism faces often vertically striated or grooved; crystals can be very large, more than 200 tonnes; also in columnar aggregates.

Physical properties. Cleavage indistinct on $\{0001\}$, fracture conchoidal or uneven. Hardness $7\frac{1}{2}$–8; density about 2.7. Colour is mostly pale green or blue, more rarely colourless,

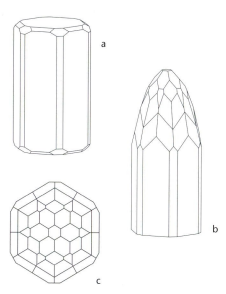

Figure 435. Beryl: (a) $\{10\bar{1}0\}$, $\{11\bar{2}0\}$, $\{0001\}$, $\{10\bar{1}2\}$, and $\{11\bar{2}2\}$; (b) $\{10\bar{1}0\}$, $\{11\bar{2}0\}$, $\{0001\}$, $\{31\bar{4}1\}$, $\{21\bar{2}1\}$, $\{20\bar{2}1\}$, $\{11\bar{2}1\}$, and $\{10\bar{1}1\}$; (c) is (b) viewed along the c axis.

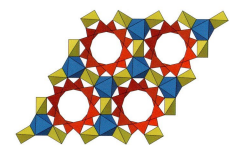

Figure 436. Sixfold rings of $[SiO_4]$ tetrahedra (red) are found in the crystal structure of beryl. These $[Si_6O_{18}]$ rings are stacked to form infinite channels along the c axis, which is perpendicular to the plane of the paper. The rings are linked by $[BeO_4]$ tetrahedra (yellow) and $[AlO_6]$ octahedra (blue).

Figure 437. Beryl var. emerald from Peru. Field of view: 50 × 75 mm.

yellow, dark green, pink, or red; vitreous lustre; transparent to translucent; large crystals can vary in transparency.

Chemical properties, etc. The crystal structure of beryl is characterized by sixfold rings of $[SiO_4]$ tetrahedra. These $[Si_6O_{18}]$ rings lie on top of each other and thus form endless channels along the c axis. The rings are linked by $[BeO_4]$ tetrahedra and $[AlO_6]$ octahedra. In addition to the main elements of beryl, minor amounts of Li, Na, or other alkali metals can be present, which like H_2O and CO_2 can be accommodated in the channels. Small amounts of Cr^{3+}, Mn^{2+}, or other transition elements give rise to various colours in beryl, e.g. Cr^{3+} is responsible for the characteristic deep green colour of emerald.

Names and varieties. Beryl is an important gemstone and is found in a variety of colours. *Aquamarine* is light green or blue; *emerald* is deep green; *morganite* or *rose beryl* is pink to deep rose; *red beryl* is darker red, and *golden beryl* or *heliodor* is golden-yellow.

Occurrence. Beryl is a common mineral primarily occurring in granites and granite pegmatites. It is also found in mica-schists and in veins and cavities in bituminous lime-

Figure 438. Beryl var. aquamarine from Mimoso do Sul, Espirito Santo, Brazil. Subject: 50 × 85 mm.

stones, e.g. in the famous emerald mines at Muzo in Colombia. The largest deposits of gemstone beryl are found in Minas Gerais, Brazil.

Use. Beryl is an important gemstone, and emerald ranks as one of the most precious gems. Beryl is the primary source of beryllium, a metal that because of its very low density is highly useful in various alloys, especially with copper. Be is also used in the nuclear power industry, e.g. as a neutron reflector.

Diagnostic features. Crystal habit, hardness, colour (in part), and (generally) mineral association and mode of occurrence. Apatite sometimes resembles beryl but has inferior hardness.

Figure 440. Dioptase: {11$\bar{2}$0}, {02$\bar{2}$1}, and {13$\bar{4}$1}.

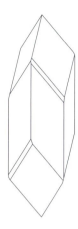

Figure 439. Red beryl from Thomas Range, Utah, USA. Field of view: 19 × 25 mm.

Dioptase, $CuSiO_3 \cdot H_2O$

Trigonal, occurs as prismatic crystals terminated by rhombohedral faces. Cleavage good on {10$\bar{1}$1}; hardness 5, density 3.3. Colour is emerald-green; vitreous lustre; transparent to translucent. Dioptase is found in the oxidized zones of Cu ores, commonly associated with malachite. It is occasionally seen as a gemstone.

Cordierite, $Mg_2Al_4Si_5O_{18}$

Crystallography. Orthorhombic $2/m2/m-2/m$; crystals normally short prismatic, commonly pseudo-hexagonal as a result of twinning on {110}; mostly as irregular grains or masses.

Physical properties. Cleavage indistinct on {010}, fracture conchoidal or uneven. Hardness 7–7½, density about 2.6. Colour is light to dark blue or violet, also colourless or grey, distinct pleochroism; vitreous lustre; transparent or translucent.

Chemical properties, etc. To a limited extent Mg can be replaced by Fe or Mn. Cordierite has sixfold rings like beryl, but in cordierite the rings are composed of four [SiO_4] tetrahedra and two [AlO_4] tetrahedra. As a result, the symmetry is reduced from hexagonal to orthorhombic. The rings form chan-

Figure 441. Dioptase from Tsumeb, Namibia. Field of view: 15 × 21 mm.

nels and, as in the case of beryl, H_2O, Na, and other alkali metals can be present in these channels.

Names and varieties. *Iolite* and *dichroite* are obsolete names for cordierite. Dichroite re-

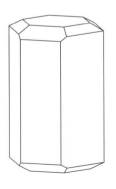

Figure 442. Cordierite: {100}, {001}, {110}, {101}, and {112}.

fers to the variation in colour with crystallographic orientation.

Occurrence. Cordierite is a characteristic mineral in contact- or regionally metamorphosed rocks rich in Al, such as hornfels, mica-schists and gneisses. Cordierite is known from, e.g. Orijärvi in Finland, Kragerø in Norway, Haddam in Connecticut, USA, and in gem quality from Mt. Bity, Madagascar. Cordierite is readily altered to tabular or dense masses of micaceous or chlorite-like minerals.

Use. Transparent varieties of cordierite, e.g. from Sri Lanka, are occasionally seen as gemstones.

Diagnostic features. When the colour is typically blue and the pleochroism distinct, cordierite is easily recognized; it otherwise resembles quartz.

292

Tourmaline,

$$AB_3C_6(BO_3)_3Si_6O_{18}(OH)_4$$

In the general formula, A is Na, Ca, or K; B is Al, Fe, Li, Mg, Mn, or Ti; C is Al, Cr, Fe, or V.

Tourmaline is a boron-containing silicate typically occurring in granite pegmatites. It is in fact a group of minerals, for the chemical composition varies considerably. Well-developed crystals of tourmaline are identified both by the characteristic crystal habit and, often, by the variation in colour within a single crystal.

Crystallography. Trigonal $3m$; crystals common, commonly dominated by the prisms $\{10\bar{1}0\}$ or $\{01\bar{1}0\}$ and with narrow faces of a hexagonal prism; prism faces commonly vertically striated; crystals are hemimorphic, i.e. they lack a centre of symmetry; the terminating faces of the two ends of the c axis consequently do not belong to the same crystal forms; the hemimorphic nature or polarity is also reflected in the physical properties; also found massive or in columnar or radiating aggregates.

Physical properties. No distinct cleavage, fracture conchoidal or uneven. Hardness 7–7½, density 3.0–3.2. Colour varies with the chemical composition: usually black, less frequently brown, green, pink, blue, yellow, or colourless; individual crystals can be multicoloured with variation in colour both along and across the c axis; vitreous lustre; transparent to translucent. Tourmaline is strongly pyro- and piezoelectric.

Figure 443. Cordierite in quartz from Tvedestrand, Norway. Subject: 49 × 71 mm.

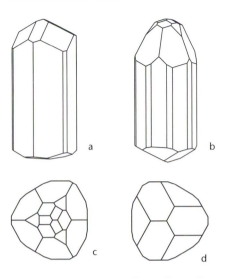

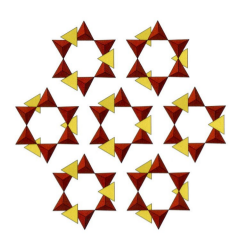

Figure 444. Tourmaline: (a) {01$\bar{1}$0}, {11$\bar{2}$0}, {02$\bar{2}$1}, {01$\bar{1}\bar{1}$}, {10$\bar{1}$1}, {10$\bar{1}$2}, and {000$\bar{1}$}; (b) {10$\bar{1}$0}, {01$\bar{1}$0}, {11$\bar{2}$0}, {41$^-$50}, {10$\bar{1}$1}, {02$\bar{2}$1}, {01$\bar{1}$2}, {32$\bar{5}$1}, {$\bar{1}$01$\bar{1}$}, and {0$\bar{2}$2$\bar{1}$}; (c) is (b) viewed from above and (d) is (b) viewed from below.

Figure 445. Triangular [BO$_3$] groups (yellow) and six-fold rings of [SiO$_4$] tetrahedra (red) are present in the crystal structure of tourmaline. The sixfold rings are polar, i.e. they all have their tetrahedra pointing to the same end of the c axis. The rings are viewed along the c axis; they are not located in one plane but at three different levels.

Chemical properties, etc. Considerable chemical variation results in 14 different end-members, of which the best known are the black *schorl*, NaFe$_3$Al$_6$(BO$_3$)$_3$Si$_6$O$_{18}$(OH)$_4$, the usually brown *dravite*, NaMg$_3$Al$_6$-(BO$_3$)$_3$Si$_6$O$_{18}$(OH)$_4$, and *elbaite*, Na(Li,Al)$_3$Al$_6$-(BO$_3$)$_3$Si$_6$O$_{18}$(OH)$_4$, which is mostly green but can have many different colours. The structural framework as illustrated in Figure 445 is chemically stable. It is composed of triangular [BO$_3$] groups and sixfold rings of [SiO$_4$] tetrahedra. All the tetrahedra are oriented with an apex pointing in the same direction, i.e. they are polar.

Names and varieties. In addition to the names of the chemical end-members, a number of gemstone varieties are associated with tourmaline: e.g. *rubellite*, which is pink to

Figure 446. Schorl (tourmaline group), from Ramfoss, Snarum, Norway. Field of view: 66 × 93 mm.

Figure 447. A slice of tourmaline from Teofilo Otoni, Minas Gerais, Brazil. Diameter: 28 mm.

red, *verdelite*, which is green in various shades, and the rarer *indigolite*, which is blue. **Occurrence.** Tourmaline occurs in granite pegmatites and frequently also in adjacent rocks that have been affected by the pneumatolytic processes that form tourmaline. It is also found as an accessory mineral in igneous rocks and many metamorphic rocks, e.g. gneisses and crystalline limestones. Schorl is the most common tourmaline; it is typically found together with the common pegmatitic minerals feldspar, quartz, and muscovite. The lighter-coloured tourmalines are commonly associated with beryl, lepidolite, fluorite, and apatite. Tourmaline localities are numerous. Most tourmalines of gemstone quality come from Minas Gerais in Brazil, others from Namibia, Madagascar, and Sri Lanka. Among

Figure 448. Tourmaline from Himalaya Mine, California, USA. Subject: 22 × 46 mm.

Figure 449. Steenstrupine from the Ilímaussaq complex, Greenland. Field of view: 13 × 21 mm.

the classic localities are Elba, Italy, and Mursinka and other places in the Urals, Russia. Beautiful crystals are known from the Himilaya mine and other mines in the Pala and Mesa Grande districts of California, USA.

Use. Tourmaline is among the most popular gemstones.

Diagnostic features. Crystal habit, including the different terminations of the crystal, the trigonal cross-section of crystals, the zoned colour features, and the lack of cleavage in comparison with, e.g. the black amphiboles.

Other cyclosilicates

Steenstrupine,
$Na_{14}Ce_6Mn_2Fe_2Zr(PO_4)_7Si_{12}O_{36}(OH)_2 \cdot 3H_2O$

Trigonal; usually found as tabular crystals, more or less metamict owing to a minor content of Th. No distinct cleavage, fracture uneven; hardness 4–5; density about 3.4. Colour is dark brown to black; submetallic lustre; almost opaque. Steenstrupine is a characteristic mineral in some veins and pegmatites in nepheline-syenites and sodalite-syenites, e.g. in the Ilímaussaq complex, Greenland.

Milarite, $(K,Na)Ca_2(Be,Al)_3Si_{12}O_{30} \cdot H_2O$

Hexagonal and mostly found in prismatic crystals. Fracture conchoidal; hardness 6;

density 2.5. Colourless to light green or yellow. Milarite occurs in veins and cavities in granite and similar rocks. *Osumilite*, $K(Fe,Mg)_2(Al,Fe)_3(Si,Al)_{12}O_{30}$, is a related mineral found in druses in volcanic rocks.

Eudialyte,
$Na_{15}Ca_6Fe_3Zr_3Si(Si_{25}O_{73})(O,OH,H_2O)_3$
$(Cl,OH)_2$

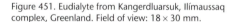

Trigonal; found as tabular or rhombohedral crystals, but mostly granular. No distinct cleavage; hardness about 5½; density 2.9. Colour is pink, red to brown, rarely yellowish; vitreous lustre. Mn, rare-earth elements, and other elements can be minor components. Eudialyte is a rock-forming mineral in some nepheline-syenites and may locally occur in large amounts. It is known from, e.g. the Ilímaussaq complex, Narssârssuk, from the great Khibina and Lozovero massifs in the Kola Peninsula, Russia, and Mont Saint-Hilaire, Quebec, Canada.

Figure 450. Milarite from Graubünden, Switzerland. Field of view: 22 × 29 mm.

Figure 451. Eudialyte from Kangerdluarsuk, Ilímaussaq complex, Greenland. Field of view: 18 × 30 mm.

Inosilicates

In inosilicates the $[SiO_4]$ tetrahedra form infinite chains. The chains can be simple single chains, as in pyroxenes; less simple single chains, as in wollastonite; and double chains, as in amphiboles. In single chains the $[SiO_4]$ tetrahedra share two corners with others, resulting in an Si/O ratio of 1:3, exactly as in cyclosilicates. In the double chains of amphiboles, half the $[SiO_4]$ tetrahedra share two corners and alternate with the other half sharing three corners, which gives an Si/O ratio of 4:11. In general, inosilicates have a prismatic, radiating, or fibrous habit along the direction of the chains. The direction and width of the chains are reflected in the cleavage properties, which are therefore among the most significant characteristics of the inosilicates.

The pyroxene group

The pyroxenes are widespread in igneous and metamorphic rocks. They are formed at higher temperatures than amphiboles, and in contrast to them do not contain OH groups. In other respects pyroxenes and amphiboles have many common features, structurally, chemically, and physically. They can be distinguished by the angles between the two cleavage directions, which in pyroxenes is 93° or 87°, and in amphiboles 124° or 56°.

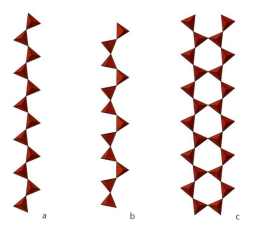

Figure 453. Chains of $[SiO_4]$ tetrahedra: (a) single chain, as in pyroxenes; (b) single chain as in wollastonite; (c) double chain as in amphiboles.

Most pyroxenes are monoclinic, but a minor group is orthorhombic. The structure of pyroxenes is characterized by single chains of $[SiO_4]$ tetrahedra, giving an Si/O-ratio of 1:3 in the chemical formula. An example of the crystal structure in pyroxenes is presented under diopside. The general formula is $XYSi_2O_6$, in which X is Ca^{2+}, Na^+, Mg^{2+}, Fe^{2+}, Mn^{2+}, or Li^+, and Y is Mg^{2+}, Fe^{2+}, Mn^{2+}, Fe^{3+}, Al^{3+}, Cr^{3+}, or Ti^{4+}.

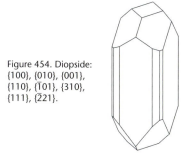

Figure 454. Diopside: {100}, {010}, {001}, {110}, {$\overline{1}$01}, {310}, {111}, {$\overline{2}$21}.

Figure 452. Diopside from Skardu, Pakistan. Field of view: 23 × 40 mm.

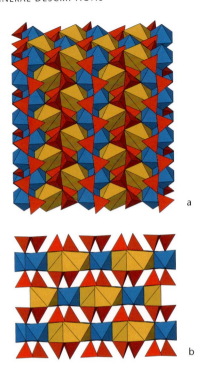

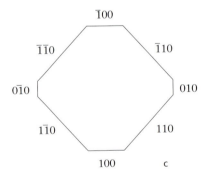

Figure 456. Diopside from Richville, New York, USA. Subject: 15×26 mm.

Figure 455. Infinite single chains of [SiO$_4$] tetrahedra (red) are present along the c axis in the crystal structure of diopside. These are linked by [(Mg,Fe)O$_6$] octahedra (blue) and Ca in irregular polyhedra with eight corners (yellow). In (a) the c axis is N–S in the plane of the paper and in (b) and (c) perpendicular to it; in all cases the b axis is E–W. By comparing (b) and (c) it is apparent that the prominent cleavage directions after the prism {110} that are typical of pyroxenes are inherent in the crystal structure. (Compare with Figure 468.)

Diopside, CaMgSi$_2$O$_6$– hedenbergite, CaFeSi$_2$O$_6$

Crystallography. Monoclinic $2/m$; diopside chiefly found as prismatic crystals, usually in combinations of {100}, {010}, and {110}, occasionally with subordinate pyramids; angles between faces belonging to the {110} prism are about 87 and 93°, i.e. almost at right angles; twinning on {100} common, often repeated; twinning on {001} less common; mostly in granular, massive, or columnar aggregates.

Physical properties. Cleavage distinct on {110}, i.e. in two directions nearly at right angles; also parting on {100} or {001}; uneven fracture. Hardness 6, density about 3.3. Colour is white to light green in diopside, darkening to nearly black with increasing Fe content; vitreous lustre; transparent or translucent.

Chemical properties, etc. Diopside and hedenbergite form a complete solid-solution series, i.e. Mg^{2+} and Fe^{2+} can substitute for each other in all proportions. The crystal structure of diopside, which is essentially the same for all pyroxenes, is shown in Figure 455. Infinite single chains of $[SiO_4]$ tetrahedra parallel to the c axis are linked by $[(Mg,Fe)O_6]$ octahedra and Ca in irregular polyhedra with eight corners. A view of the crystal structure seen along the c axis shows how the cleavage on {110} can take place without breaking any chains of $[SiO_4]$ tetrahedra.

Names and varieties. *Augite* is a closely related mineral mentioned below. *Chrome diopside* is a Cr-bearing variety of diopside of emerald-green colour found, e.g. at Outokumpu, Finland. *Johannsenite*, $CaMnSi_2O_6$, forms a complete solid-solution series with hedenbergite.

Occurrence. Diopside and hedenbergite occur typically in Ca-rich metamorphic rocks of medium to high grade. Diopside is also a characteristic mineral in contact-metamorphosed limestones; hedenbergite is characteristic of the corresponding Fe-rich rocks. Both diopside and hedenbergite are also

Figure 457. Hedenbergite from Nordmark, Värmland, Sweden. Field of view: 26 × 37 mm.

Figure 458. Diopside with parting on {001} from Outokumpu, Finland. Subject: 60 × 66 mm.

301

found in basic igneous rocks. Fine crystals of diopside are found, e.g. at De Kalb and other places in New York State, USA, and of hedenbergite, e.g. at Nordmark, Sweden.
Use. Diopside is occasionally seen as a gemstone.
Diagnostic features. Crystal habit and, in particular, cleavage.

Augite, $(Ca,Na)(Mg,Fe,Al)(Si,Al)_2O_6$

Augite is closely related to the diopside–hedenbergite series but differs by having (Mg,Fe) and Si partially replaced by Al. Small amounts of Ca are often replaced by Na, and in addition Ti can to a limited extent replace (Mg,Fe). The crystallographic and physical properties

Figure 461. Augite, twinned on {100}, from Schima, Bohemia, Czech Republic.
Subject: 15×33 mm.

are the same as for diopside–hedenbergite except that augite is usually dark green to black. Augite is the most common pyroxene and one of the principal minerals in igneous rocks, particularly in basic and ultrabasic rocks such as syenites, gabbros, basalts, and pyroxenites. Fine crystals are known, e.g. from the lavas of Vesuvius, Italy. *Omphacite*, $(Ca,Na)(Mg,Fe,Al)$-

Figure 459. Augite: {100}, {110}, {010}, and {$\bar{1}$11}.

Figure 460. Augite from Arendal, Norway. Subject: 43×78 mm.

Si_2O_6, is a related green mineral, which together with garnet constitutes the rock *eclogite*. Augite is recognized by its crystal habit, prismatic cleavage, and colour.

Pigeonite, $(Mg,Fe,Ca)_2Si_2O_6$

Pigeonite is as a Ca-poor augite formed at high temperature; it occurs primarily in basalts and other rapidly cooled igneous rocks. Pigeonite is distinguished from other pyroxenes by its optical properties.

Aegirine, $NaFe^{3+}Si_2O_6$

Crystallography. Monoclinic $2/m$; crystals commonly long prismatic with more-or-less steep terminations; also needle-shaped or fibrous; twinning on {100} common; also granular.

Physical properties. Cleavage distinct on {110}. Hardness 6, density about 3.5. Colour is dark green to greenish-black or brownish-black; streak is yellowish-green in contrast to the otherwise similar arfvedsonite with a bluish-grey streak; vitreous lustre; translucent.

Chemical properties, etc. There is a solid-solution series between aegirine and augite, $(Ca,Na)(Mg,Fe,Al)(Si,Al)_2O_6$. $NaFe^{3+}$ in aegirine is in general replaced to a moderate extent by $CaFe^{2+}$ in particular. Aegirine has a diopside crystal structure.

Names and varieties. *Acmite* is a synonym for aegirine but has also been used for brown aegirine.

Occurrence. Aegirine is found in alkali-rich igneous rocks such as nepheline-syenites, alkali granites, and alkali syenites with asso-

Figure 462. Aegirine: {100}, {110}, {310}, {221}, and {661}.

Figure 463. Aegirine, radiated, from Narssârssuk, Greenland.
Subject: 86 × 99 mm.

ciated pegmatites. Fine crystals are known, e.g. from Narssârssuk in Greenland, the Lovozero and Khibina massifs in the Kola Peninsula, Russia, Zomba-Malosi in Malawi, and Mont Saint-Hilaire, Quebec, Canada.

Figure 464. Aegirine from Narssârssuk, Greenland. Subject: 11 × 86 mm.

Diagnostic features. Crystal habit, colour, mineral association, streak, and cleavage, the last two especially when compared with the otherwise similar amphibole arfvedsonite.

Jadeite, $Na(Al,Fe)Si_2O_6$

Crystallography. Monoclinic $2/m$; crystals rare, commonly granular, massive, fibrous.

Physical properties. Cleavage as for other pyroxenes but rarely visible; fine-grained masses very tough. Hardness $6\frac{1}{2}$, density about 3.3. Colour is light to dark green, rarely white or brown; vitreous lustre, slightly greasy on polished surfaces; translucent.

Chemical properties, etc. The replacement of Al by Fe^{3+} is limited. Jadeite has a diopside crystal structure. Its notable toughness is due to a microcrystalline aggregate texture.

Names and varieties. Jadeite is named after *jade*, a common name for jadeite and the amphibole nephrite, which has practically the same physical properties.

Occurrence. Jadeite is formed in metamorphic rocks under high pressure and at low temperature. First recognized as boulders in river beds, it is in particular known from deposits in Myanmar (Burma) and other deposits in South-east Asia.

Use. Jade, i.e. jadeite and nephrite, is a highly appreciated material for carving. In early times it was also used for tools.

Diagnostic features. Colour and toughness; difficult to distinguish from nephrite, which has a slightly lower density.

Spodumene, $LiAlSi_2O_6$

Crystallography. Monoclinic $2/m$; crystals common, commonly long prismatic, flattened parallel to {100} and with marked striation parallel to the c axis; crystals can be large, up to several tonnes; twinning on {100} common.

Physical properties. Cleavage perfect on {110}, parting usually distinct on {100}. Hard-

Figure 465. Jadeite from Clear Creek, San Benito County, California, USA. Subject: 37 × 44 mm.

ness 6½–7, density about 3.2. Colour is white or grey, rarely pink, green, or yellow; vitreous lustre; transparent or translucent.

Chemical properties, etc. Spodumene has a diopside crystal structure.

Names and varieties. Spodumene of gemstone quality is known as *kunzite* when violet, lilac, or pink, and as *hiddenite* when green.

Occurrence. Spodumene is found almost exclusively in Li-rich granite pegmatites. It is known, e.g. from Pala Chief mine and other mines in the Pala District, California, the Etta mine in South Dakota, USA, and the Urupuca and other mines in Minas Gerais, Brazil.

Figure 466. Spodumene, Urupuca Mine, Itambacuri, Minas Gerais, Brazil. Subject: 55 × 132 mm.

Figure 467. Enstatite from Bamble, Norway.
Subject: 80 × 128 mm.

Use. Spodumene is an appreciated gemstone and has been an important source of lithium.
Diagnostic features. Crystal habit, cleavage, colour, and mode of occurrence.

Enstatite, (Mg,Fe)SiO₃–ferrosilite, (Fe,Mg)SiO₃

Enstatite, $(Mg,Fe)SiO_3$–
ferrosilite, $(Fe,Mg)SiO_3$

Crystallography. Orthorhombic $2/m2/m2/m$; crystals rare, mostly granular, fibrous.
Physical properties. Cleavage distinct on {210}, i.e. with the same angles between cleavage directions as in monoclinic pyroxene. Hardness 5–6, density about 3.2, increasing with Fe content. Colour is grey, green, or brown, darkening to black with increasing Fe content; vitreous lustre, pearly on cleavage surfaces; some compositions can have a sub-metallic bronze-like lustre; translucent.
Chemical properties, etc. Enstatite and ferrosilite form an almost complete solid-solution series. The crystal structure of enstatite can be described as a twinning of the diopside structure on {100}, which results in a doubling of the *a* axis. This explains why the same prismatic cleavage is called {110} in monoclinic pyroxenes and {210} in orthorhombic pyroxenes.
Names and varieties. *Bronzite* and *hypersthene* are older names for intervals within this solid-solution series.
Occurrence. Enstatite and the Mg-rich part of the series are common minerals in basic and ultrabasic rocks such as basalts, norites, peridotites, and pyroxenites. Pure ferrosilite is rare in nature.
Diagnostic features. The prismatic cleavage and the peculiar lustre sometimes seen are of some help, otherwise these minerals, especially the Fe-rich members, are usually difficult to identify.

The amphibole group

The large amphibole group includes a series of important minerals that occur in igneous and metamorphic rocks. In general they form at somewhat lower temperatures than pyroxenes and in conditions where OH is available. Amphiboles resemble pyroxenes in many ways, structurally as well as in chemical and physical properties, but in chemical composition they are far more complicated. The principal visual difference between amphiboles and pyroxenes is the angles between the two cleavage directions, which in amphiboles is 124° or 56° and in pyroxenes 93° or 87°.

Amphiboles are monoclinic or ortho-rhombic. The crystal structure of amphiboles is characterized by double chains of [SiO$_4$] tetrahedra, giving an Si/O-ratio of 4:11 in the general formula. Further details of the structure are given under tremolite. The general formula for amphiboles is $WX_2Y_5Si_8O_{22}$-(OH)$_2$, in which W is Na$^+$, K$^+$, or (mostly) nothing; X is Ca^{2+}, Na$^+$, Mg^{2+}, Fe^{2+}, Mn^{2+}, or Li$^+$; and Y is Mg^{2+}, Fe^{2+}, Mn^{2+}, Fe^{3+}, Al^{3+}, or Ti^{4+}. Si^{4+} can to a larger extent than in pyroxenes be replaced by Al^{3+}.

Tremolite, $Ca_2(Mg,Fe)_5Si_8O_{22}(OH)_2$– ferro-actinolite $Ca_2(Fe,Mg)_5Si_8O_{22}(OH)_2$

Crystallography. Monoclinic $2/m$; crystals commonly long prismatic in radiating or columnar aggregates, also fibrous or asbestiform and in microcrystalline dense masses; twinning on {100} seen.

Physical properties. Cleavage perfect on {110}, i.e. two directions with 124° or 56° between them; dense fine-grained masses; tough. Hardness 5½–6; density 3.0–3.4, increasing with Fe content. Colour is white, grey, light green to dark green, or nearly black, darkening with increasing Fe content; vitreous lustre; transparent or translucent.

Chemical properties, etc. There is a complete solid-solution series between tremolite and ferro-actinolite; the name *actinolite* is used for intermediate compositions with Mg > Fe. The crystal structure of tremolite, which is essentially the same for all amphiboles, is shown in Figure 468. Infinite double chains of [SiO$_4$] tetrahedra parallel to the c axis are linked by [(Mg,Fe)O$_6$] octahedra and Ca in irregular polyhedra with eight corners.

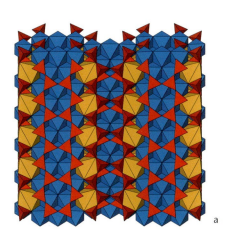

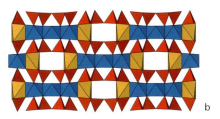

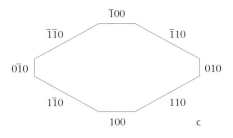

Figure 468. Infinite double chains of [SiO$_4$] tetrahedra (red) along the c axis are present in the crystal structure of tremolite. They are linked together by [(Mg,Fe)O$_6$] octahedra (blue) and by Ca in irregular polyhedra with eight corners (yellow). In (a) the c axis is N–S in the plane of the paper; in (b) and (c) the c axis is perpendicular to it. In all three cases the b axis is E–W. By comparing (b) and (c) it is apparent that the two prominent directions of cleavage after the prism {110} typical of amphiboles is inherent in the crystal structure. (Compare with Figure 455.)

307

A view of the crystal structure seen along the *c* axis shows how the cleavage on {110} can take place without breaking any double chains of [SiO₄] tetrahedra.

Names and varieties. *Nephrite* is a microcrystalline tough variety that together with jadeite is known by the common name *jade*. Nephrite closely resembles jadeite but has a lower density.

Occurrence. The members of the series are common minerals in low- and medium-grade metamorphic rocks; tremolite in particular is a characteristic mineral in metamorphosed dolomitic limestones. Val Tremola, St. Gotthard in Switzerland is among the countless localities notable for fine crystals.

Use. Asbestiform tremolite has been of some importance as a crude ore for asbestos. Like jadeite, nephrite is used for carving and was also in early times used for axes and other tools, as, e.g. by the Maori people in New Zealand.

Diagnostic features. The columnar or fibrous aggregate forms, cleavage, and, usually, the light colour.

Figure 470. Tremolite from Tirol, Austria. Subject: 51 × 76 mm.

Figure 469. Actinolite from Tirol, Austria. Subject: 74 × 170 mm.

Figure 471. Actinolite var. nephrite from Westland, South Island, New Zealand. Face of specimen polished. Subject: 124 × 132 mm.

Hornblende, e.g.
$Ca_2(Fe^{2+},Mg)_4(Al,Fe^{3+})(Si_7Al)O_{22}(OH,F)_2$

Hornblende is a group of minerals with relationships to the tremolite–ferro–actinolite series that are similar to those between augite and the diopside–hedenbergite series. Two end-members are recognized: *ferrohornblende*, and *magnesiohornblende*. The hornblendes are characterized by a great variation in chemical composition, primarily in higher contents of Al, Fe^{3+}, and Na in comparison with tremolite–ferro–actinolite. The colour is dark green to black. The hornblendes share most physical properties with the tremolite–ferro–actinolite series, except that they are not fibrous. Hornblendes are important rock-forming minerals, especially as essential constituents of medium-grade metamorphic rocks such as hornblende-schists and amphibolites. They are also found in igneous rocks such as basalts, granites, and syenites, etc. and in related pegmatites.

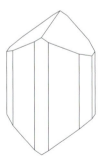

Figure 472. Hornblende: {100}, {010}, {110}, {120}, and {021}

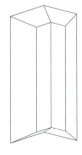

Figure 473. Hornblende: Twinned on {100} with {010}, {110}, and {011}.

Figure 474. Hornblende from Nordmark, Värmland, Sweden. Subject: 75 × 84 mm.

Edenite, pargasite, hastingsite, tschermakite, and *kaersutite* are all amphiboles closely related to hornblende.

Hornblende is recognized by its crystal habit, prismatic cleavage, and nearly black colour.

Glaucophane,
$Na_2(Mg,Fe)_3Al_2Si_8O_{22}(OH)_2$

Crystallography. Monoclinic $2/m$; crystals long prismatic or needle-shaped, commonly in radiating or granular aggregates.

Physical properties. Cleavage distinct on {110}. Hardness about 6, density about 3.0–3.2, increasing with Fe content. Colour is bluish-grey, blue to almost black; vitreous lustre; translucent.

Figure 475. Pargasite from Ersby, Pargas, Finland. Field of view: 36 × 50 mm.

Figure 476. Kaersutite from Qaarsut (formerly Qaersut), Greenland. Subject: 131 × 166 mm.

Chemical properties, etc. Fe^{3+} can in small amounts replace Al. Glaucophane is strictly an end-member of a glaucophane–ferroglaucophane series.

Occurrence. Glaucophane is found in metamorphic rocks, especially those formed at low temperature and relatively high pressure. It is a main constituent of glaucophane-schists.

Diagnostic features. Colour and mode of occurrence.

Figure 477. Glaucophane from Val di Susa, Torino, Italy. Subject: 61 × 149 mm.

Figure 478. Riebeckite var. crocidolite from Griqualand East, Republic of South Africa. Field of view: 40×60 mm.

Riebeckite,
$$Na_2(Fe^{2+},Mg)_3(Fe^{3+})_2Si_8O_{22}(OH,F)_2$$

Crystallography. Monoclinic $2/m$; mostly in radiating, felted, or asbestiform aggregates.

Physical properties. Cleavage perfect on {110}. Hardness 6, density about 3.4. Colour is blue or bluish-black; vitreous lustre, silky when fibrous; translucent.

Chemical properties, etc. Mg^{2+} can predominate over Fe^{2+}. Riebeckite is strictly an end-member of a riebeckite–magnesioriebeckite series.

Names and varieties. *Crocidolite* or 'blue asbestos' is an asbestiform variety of riebeckite. *Tiger's eye* is a gemstone, in which crocidolite is replaced by quartz while the fibrous structure is preserved, which results in a silky-shining brown coloured stone with a chatoyancy (a lustre resembling the eye of a cat).

Occurrence. Riebeckite primarily occurs in alkali granites, syenites and nepheline-syenites, and related pegmatites. The largest deposits of crocidolite have been found in the Cape Province, Republic of South Africa, where they have been important asbestos ores.

Diagnostic features. Colour and mode of occurrence.

Arfvedsonite,
$$Na_3(Fe^{2+},Mg)_4Fe^{3+}Si_8O_{22}(OH)_2$$

Crystallography. Monoclinic $2/m$; found in long prismatic crystals, commonly flattened parallel to {100}, mostly in columnar or granular aggregates; twinning on {100} seen.

Physical properties. Cleavage perfect on {110}. Hardness 5½–6, density about 3.4. Colour is bluish-black to black; streak bluish-

Figure 479. Arfvedsonite from Kangerdluarsuk, Ilímaussaq complex, Greenland. Subject: 97×147 mm.

Figure 480. Richterite from Långban, Värmland, Sweden. Field of view: 32 × 45 mm.

grey, in contrast to aegirine, which is otherwise similar but has a yellowish-green streak; vitreous lustre; translucent.

Chemical properties, etc. Al^{3+} can to some extent replace Fe^{3+}, and Mg^{2+} sometimes predominates over Fe^{2+}. Arfvedsonite forms a series with *magnesio-arfvedsonite*.

Names and varieties. *Katophorite and eckermannite* are related amphiboles with modes of occurrence similar to those of arfvedsonite; in contrast, *richterite*, also related to arfvedsonite, is found in contact-metamorphosed limestones or Mn-rich ore deposits, as, e.g. at Långban, Sweden.

Occurrence. Arfvedsonite occurs in alkaline complexes, where it is found as rock-forming mineral as well as in larger crystals in pegmatites. It is known, e.g. from Ilímaussaq in Greenland, Langesundsfjorden, Norway, and Mont Saint-Hilaire, Quebec, Canada..

Diagnostic features. Crystal habit, cleavage, colour, streak, and mode of occurrence.

Cummingtonite,
$(Mg,Fe,Mn)_7Si_8O_{22}(OH)_2$

Crystallography. Monoclinic $2/m$; crystals uncommon, typically in needle-shaped or fibrous aggregates, often radiating.

Physical properties. Cleavage perfect on {110}. Hardness 6, density 3.2–3.6, increasing with Fe and Mn content. Colour is light to dark brown, sometimes green or blue; vitreous lustre, silky lustre when fibrous; translucent.

Chemical properties. Cummingtonite forms a solid-solution series with *grunerite*, $(Fe,Mg)_7$-$Si_8O_{22}(OH)_2$.

Names and varieties. *Amosite* is a trade name for asbestiform amphibole varieties, mainly in the cummingtonite–grünerite series.

Occurrence. Cummingtonite is found in regional metamorphic rocks such as amphibo-

Figure 481. Cummingtonite var. amosite from Penge, Northern Province, Republic of South Africa. Subject: 90 × 108 mm.

lites, in which it commonly occurs in association with other amphiboles.

Diagnostic features. Colour and radiating aggregate form; difficult to distinguish from anthophyllite.

Anthophyllite, $(Mg,Fe)_7Si_8O_{22}(OH)_2$

Crystallography. Orthorhombic $2/m2/m2/m$; crystals uncommon, typically as lamellar or fibrous aggregates.

Physical properties. Cleavage perfect on {210}, i.e. two cleavage directions with approximately the same angles as in monoclinic amphiboles. Hardness 6; density 2.8–3.3, increasing with Fe content. Colour is brown, yellowish or greyish, rarely greenish; vitreous to silky lustre; transparent to translucent.

Chemical properties, etc. Anthophyllite forms a solid-solution series with *gedrite*, $(Mg,Fe)_5Al_2(Si_6Al_2)O_{22}(OH)_2$, an Al-bearing orthorhombic amphibole.

315

Figure 482. Anthophyllite from Kjennerudvann Quarry, Kongsberg, Norway. Subject: 81 × 114 mm.

Names and varieties. Gedrite is known, e.g. from Snarum, Norway, and Qeqertarsuatsiaat (Fiskenæsset), Greenland. *Holmquistite* is a closely related Li-bearing amphibole first described from Utö, Sweden.

Occurrence. Anthophyllite occurs in Mg-rich metamorphic rocks of medium grade, typically in association with talc or cordierite. It is known, e.g. from Kongsberg in Norway, and several places in Pennsylvania, USA.

Use. Asbestiform anthophyllite has been used in the asbestos industry.

Diagnostic features. The brown colour can be characteristic; otherwise anthophyllite is difficult to distinguish from, e.g. cummingtonite.

Figure 483. Holmquistite from Utö, Södermanland, Sweden. Subject: 55 × 70 mm.

Other inosilicates

Besides the two large groups of inosilicates, pyroxenes and amphiboles, a great number of inosilicates exist that have more or less complicated structures. Of these, a few such as wollastonite are fairly abundant, whereas the majority of the others minerals are of importance only within restricted regions.

Wollastonite, $CaSiO_3$

Crystallography. Triclinic $\bar{1}$; good crystals rare, mostly tabular dominated by {100} and {001} and elongated along the b axis; twin-ning on {100} seen; mostly in fibrous, radiating, columnar aggregates.

Physical properties. Cleavage perfect on {100} and {001}, good on {$\bar{1}$01}, giving splinters parallel to the b axis. Hardness 5; density 2.9. Colourless, white or grey; vitreous lustre, silky when fibrous; translucent.

Chemical properties, etc. To some extent Ca can be replaced by Fe or Mn. The crystal structure of wollastonite consists of infinite single chains along the c axis linked by irregular [CaO_6] octahedra. The linkage of chains and octahedra is arranged in various ways, giving rise to different polytypes of wollastonite.

Occurrence. Wollastonite is a typical mineral in contact-metamorphic limestones, where it occurs in association with, e.g. Ca-rich garnets, diopside, tremolite, and epidote. Large

crystals are known from Diana in New York State, USA.

Use. Wollastonite is widely used in the ceramics industry. A recent use is as a substitute for the traditional asbestos minerals.

Diagnostic features. Mode of occurrence and mineral association; resembles other fibrous silicates such as tremolite and pectolite, but may be identified by the angle (about 84°) between the two best cleavage directions.

Bustamite, $CaMnSi_2O_6$

Triclinic; tabular crystals seen, but found mostly compact or fibrous. Cleavage perfect on {100}, less good on {110} and {1$\overline{1}$0}; hardness about 6; density 3.3. Colour is light red or brownish-red; vitreous lustre; translucent. Mn can be replaced in part by Fe and to a lesser extent by Zn. Bustamite occurs typically in association with metamorphic Mn-rich ore deposits and is known, e.g. from Franklin and Stirling Hill in New Jersey, USA.

Figure 484. Wollastonite from Tammela, Finland. Subject: 61×99 mm.

Figure 485. Bustamite with zincite from Franklin, New Jersey, USA. Subject: 75×103 mm.

Figure 486. Pectolite from the Batbjerg complex, Greenland. Field of view: 74 × 94 mm.

Pectolite, $NaCa_2Si_3O_8(OH)$

Crystallography. Triclinic $\bar{1}$; single crystals rare, mostly in aggregates of needle-shaped crystals elongated along the b axis; aggregates often radiating or spherical.

Physical properties. Cleavage perfect on {100} and {001}. Hardness 5; density 2.9. Colourless, white, or grey; vitreous to silky lustre; translucent.

Chemical properties, etc. The composition of pectolite is normally close to the ideal formula, but sometimes Ca is considerably replaced by Mn. Structurally, pectolite is related to wollastonite.

Names and varieties. *Schizolite* is a light red to brown Mn-bearing pectolite known, e.g. from Ilímaussaq in Greenland. *Sérandite*, $Na(Mn,Ca)_2Si_3O_8(OH)$, is a rose-red mineral with Mn > Ca; it is found, e.g. as crystals of a beautiful salmon colour in Mont Saint-Hilaire, Quebec, Canada.

Occurrence. Pectolite occurs normally in fissures and cavities in basalts, typically in association with zeolites, as at Paterson and Bergen Hill in New Jersey, USA.

Figure 487. Sérandite with analcime from Mont Saint-Hilaire, Quebec, Canada. Field of view: 34 × 39 mm.

Diagnostic features. Pectolite resembles wollastonite in physical properties as well as in aggregate form, but usually occurs in different associations.

Sorensenite, $Na_4Be_2Sn(Si_3O_9)_2\cdot 2H_2O$

Monoclinic and found as ruler-shaped crystals; two cleavage directions with 63° between them; hardness 5½; density 2.9. Colourless, white, or pink; vitreous lustre. Sorensenite is known only from the Ilímaussaq complex in Greenland, where it is relatively common in hydrothermal veins in association with, e.g. analcime, microcline, sodalite, or nepheline.

Figure 488. Sorensenite from Kvanefjeld, Ilímaussaq complex, Greenland. Field of view: 64 × 88 mm.

Rhodonite, $(Mn,Ca)_5Si_5O_{15}$

Crystallography. Triclinic $\bar{1}$; crystals typically tabular parallel to {001}, commonly with rounded edges; mostly massive, granular.

Physical properties. Cleavage perfect on {110} and {1$\bar{1}$0} at an angle close to 90°. Hardness about 6; density 3.5–3.7. Colour varies between pink and brownish-red; vitreous lustre; translucent.

Chemical properties, etc. Rhodonite is never pure $MnSiO_3$; Mn is always partly replaced by Ca and to a minor degree by Fe, and sometimes by Mg and Zn. The crystal structure of rhodonite is related to wollastonite but has a different type of single chain.

Names and varieties. *Pyroxmangite*, $Mn_7Si_7O_{21}$, resembles rhodonite chemically but has a different crystal structure. It is a reddish mineral found, e.g. in large crystals at Broken Hill in New South Wales, Australia.

Figure 489. Rhodonite from
Lângban, Värmland, Sweden.
Field of view: 56 × 82 mm.

Figure 490. Rhodonite from Broken
Hill, New South Wales, Australia.
Subject: 48 × 51 mm.

Figure 491. Babingtonite from Arendal, Norway. Subject: 37 × 63 mm.

Occurrence. Rhodonite is found in metamorphic Mn-bearing deposits and in deposits affected by hydrothermal or metasomatic processes. Large amounts are known, e.g. from Yekaterinburg in the Urals, Russia, Broken Hill, New South Wales, Australia, and Franklin and Stirling Hill in New Jersey, USA.
Use. Massive rhodonite is occasionally polished for ornamental purposes.
Diagnostic features. Crystal habit, colour, hardness, and cleavage.

Babingtonite, $Ca_2(Fe^{2+},Mn)Fe^{3+}Si_5O_{14}(OH)$

Triclinic; found as short prismatic crystals. Cleavage perfect on {110} and {1$\bar{1}$0} at an angle close to 90°. Hardness about 6; density 3.4. Colour is dark green to black; strong vitreous lustre; translucent. Babingtonite is related to rhodonite. It occurs in fissures and cavities in granitic rocks and is also known from skarn deposits. Fine crystals are found at Lane's quarry near Westfield in Massachusetts, USA.

Inesite, $Ca_2Mn_7Si_{10}O_{28}(OH)_2 \cdot 5H_2O$

Triclinic; mostly found in radiating aggregates of needle-shaped or fibrous crystals. Cleavage perfect on {010}, distinct on {100}; hardness 6; density 3.1. Colour is reddish, usually flesh-coloured; vitreous lustre; translucent. Inesite is structurally related to rhodonite. It occurs dispersed in Mn-bearing associations and is known, e.g. from Hale Creek mine in California, USA.

Chkalovite, $Na_2BeSi_2O_6$

Orthorhombic; good crystals rare. Hardness 6, density 2.7. Colourless or whitish with a somewhat greasy vitreous lustre. Chkalovite is found, e.g. in the Ilímaussaq complex in Greenland, where it occurs in ussingite veins. It is best identified by its somewhat furrowed appearance on weathered surfaces.

Lorenzenite, $Na_2Ti_2O_3(Si_2O_6)$

Orthorhombic and found mostly as prismatic crystals. Cleavage perfect on {100}, distinct on {110}; hardness 6; density 3.5. Colour is usually yellowish-brown to dark brown; vitreous lustre; transparent to nearly opaque. It is found in alkali syenites and related pegmatites, and is known, e.g. from Narssârssuk in Greenland, Mt. Flora in the Lovozero massif, Kola Peninsula, Russia, and Mont Saint-Hilaire, Quebec, Canada. *Ramsayite* is an obsolete name for lorenzenite.

Okenite, $Ca_{10}Si_{18}O_{46}\cdot18H_2O$

Triclinic; occurs mostly fibrous or needle-shaped in radiated aggregates, commonly spherical. Hardness 5; density 2.3. Colourless or white, rarely pale yellowish or bluish; pearly lustre. Okenite occurs in cavities in basalts and similar rocks and is known, e.g. from Bombay, India, Qeqertarsuaq (Disko Island), Greenland, and Skookumchuck Dam in Washington, USA.

Xonotlite, $Ca_6Si_6O_{17}(OH)_2$

Monoclinic; occurs mostly in compact, fibrous masses resembling chalcedony. Hardness about 6; density 2.7. Colour is white or grey; vitreous or pearly lustre. Xonotlite is usually found in veins in serpentinites and contact zones.

Eudidymite, $NaBeSi_3O_7(OH)$

Monoclinic and polymorph with epididymite; commonly in crystals tabular parallel to {010}; twinning on {100} frequent. Cleavage perfect on {001}, distinct on {100}; hardness about 6; density 2.6. Colourless or white; silky lustre. Eudidymite occurs in alkali syenites and related pegmatites and is

Figure 492. Inesite (red) with natrolite (white) from Wessels mine, Northern Cape Province, Republic of South Africa. Field of view: 50 × 75 mm.

Figure 493. Okenite on calcite in a geode from Poona, India. Field of view: 90 × 135 mm.

Figure 494. Epididymite from Narssârssuk, Greenland.
Subject: 36 × 43 mm.

known from, e.g. Narssârssuk in Greenland and Langesundsfjorden in Norway.

Epididymite, $NaBeSi_3O_7(OH)$

Orthorhombic and polymorphic with eudidymite; in needle-shaped, platy, or prismatic crystals, also fibrous, granular, porcelaineous; repeated twinning frequent. Cleavage perfect on {001}, distinct on {100}; hardness about 6; density 2.6. Colourless or white; silky lustre. Epididymite occurs in alkali syenites and related pegmatites and is known from, e.g. Narssârssuk, Greenland, and Langesundsfjorden, Norway.

Elpidite, $Na_2ZrSi_6O_{15} \cdot 3H_2O$

Orthorhombic; crystals commonly long prismatic parallel to [001], mostly in massive, fibrous, or long prismatic aggregates; commonly with striation parallel to the c axis. Cleavage good on {110}; hardness about 7; density about 2.6. Colourless, white, grey, yellowish, brownish, greenish, or reddish; mostly vitreous lustre. Elpidite is found in alkali granites and nepheline-syenites with associated pegmatites, and is known from a number of localities, e.g. Narssârssuk in Greenland. Crystals up to 30 cm long are found at Tarbagatai in Kazakhstan.

Astrophyllite,
$(K,Na)_3(Fe,Mn)_7Ti_2Si_8(O,OH)_{31}$

Triclinic; found mostly in tabular mica-like or needle-shaped crystals. Cleavage perfect on {001}, cleavage blades brittle; hardness 3½; density about 3.4. Brass-coloured to dark brown; submetallic vitreous lustre. Astrophyllite is found in nepheline-syenites with associated pegmatites and is known, e.g. from Langesundsfjorden in Norway, Khibina in Russia, and Mont Saint-Hilaire, Quebec, Canada.

Aenigmatite, $Na_2(Fe^{2+})_5TiSi_6O_{20}$

Triclinic; mostly in poorly developed long prism-like crystals. cleavage on {100} and {010} perfect; hardness 5½; density about 3.8. Colour is dull black or brownish-black, streak reddish-brown; vitreous lustre, slightly

Figure 495. Elpidite from Narssârssuk, Greenland. Subject: 38 × 73 mm.

Figure 496. Astrophyllite from Narssârssuk, Greenland. Field of view: 31 × 60 mm.

Figure 497. Aenigmatite in eudialyte (pink) from Kangerdluarsuk, Ilímaussaq complex, Greenland. Field of view: 40 × 58 mm.

greasy. Aenigmatite is a rock-forming mineral in some nepheline-syenites with associated pegmatites and is known, e.g. from the Ilímaussaq complex, Greenland.

Sapphirine, $Mg_7Al_{18}Si_3O_{40}$

Monoclinic or triclinic; seen in tabular crystals parallel to {010}, mostly granular. Moderate cleavage on {010}, poor on {001} and {100}; hardness 7½, density 3.5. Colour is typically sapphire-blue (hence the name), but varies into greenish nuances; vitreous lustre. Sapphirine is a characteristic mineral in high-grade metamorphic rocks rich in Mg and Al. It is known, e.g. from Qeqertarsuatsiaat (Fiskenæsset), Greenland, where it is found in association with hornblende, biotite, gedrite, and kornerupine.

Narsarsukite, $Na_2(Ti,Fe,Zr)Si_4(O,F)_{11}$

Tetragonal; mostly as tabular crystals parallel to {001}, rarely prismatic. Cleavage perfect on

Figure 498. Sapphirine from Qeqer-tarsuatsiaat (Fiskenæsset), Greenland. Subject: 147 × 163 mm.

Figure 499. Narsarsukite in quartz from Narssârssuk, Greenland. Field of view: 44 × 64 mm.

{100} and {110}; hardness 6½; density 2.8. Colour is usually honey-yellow, also dark green; vitreous lustre. Narsarsukite is found in alkaline pegmatites and is known, e.g. from Narssârssuk, Greenland, and Mont Saint-Hilaire, Quebec, Canada.

Charoite,
$(K,Na)_5(Ca,Ba,Sr)_8(Si_6O_{15})_2(Si_6O_{16})(OH,F) \cdot nH_2O$

Monoclinic; mostly in dense fibrous aggregates. cleavage on {001} good; hardness about 6; density 2.5. Colour is lilac to violet; vitreous or silky lustre. Charoite is found in metasomatically altered alkaline rocks in the Murun massif, Siberia, and is primarily known for its use in decorative carvings, vases, *en cabochons*, etc.

327

Figure 500. Neptunite (black) with benitoite (blue) on natrolite. Dallas Gem Mine, San Benito County, California, USA. Field of view: 17 × 27 mm.

Neptunite,
$KNa_2Li(Fe,Mg,Mn)_2Ti_2Si_8O_{24}$

Monoclinic; found in prismatic crystals, commonly elongated and with predominant {110}. Cleavage perfect on {110}; hardness 5½; density 3.2. Colour is black, blood-red in thin splinters; vitreous lustre. Neptunite is mostly found in association with syenitic rocks. It is known, e.g. from San Benito in California, USA, and Narssârssuk in Greenland.

Chiavennite,
$CaMn(BeOH)_2(Si,Al)_5O_{13}\cdot2H_2O$

Orthorhombic; primarily as tiny flattened spear-shaped crystals in spherical aggregates. Cleavage good on {100}, {010}, and {001}; hardness 3; density 2.6. Colour is orange-yellow, also reddish; vitreous lustre. Chiavennite is found in veins and cavities in syenitic pegmatites as well as in coatings on beryl in granite pegmatites. It is known, e.g. from Tvedalen at Larvik, Norway, and Chiavenna, Italy.

Bavenite, $Ca_4(Al,Be)_4Si_9O_{26}(OH)_2$

Orthorhombic; found in crystals of fibrous, tabular, or prismatic habit, also in dense fine-grained masses. Cleavage perfect on {001}, good on {010}; hardness 5½; density 2.7. White, colourless, or with light green or red nuances; vitreous or silky lustre. Bavenite is relatively common in small amounts and is found in cavities in granite pegmatites and pneumatolytic altered rocks. It is known, e.g. from Baveno in Piemonte, Italy, and Foote mine in North Carolina, USA

Tundrite, $Na_2Ce_2TiO_2SiO_4(CO_3)_2$

Triclinic; found in small needle-shaped crystals, commonly in radiating aggregates. Cleavage good on {010}; hardness about 3; density 3.7. Colour is yellowish-brown to yellowish-green; silky lustre. There are two species: *tundrite-(Ce)* and *tundrite-(Nd)*. The chemical formulae are also written in more complex form. Tundrite-(Ce) is found in nepheline-syenite pegmatites and is known from the Lovozero and Khibina massifs in the Kola Peninsula, Russia; tundrite-(Nd) occurs in the Ilímaussaq complex, Greenland.

Figure 501. Bavenite (light yellow) with albite (pink) on ilvaite (black) from Kangerdluarsuk, Ilímaussaq complex, Greenland. Field of view: 40 × 60 mm.

Figure 502. Tundrite from Kvanefjeld, Ilímaussaq complex, Greenland. Field of view: 23 × 36 mm.

Figure 503. Phlogopite
from Zé Pinto Prospect,
Aldeia, Minas Gerais,
Brazil.
Subject: 99 × 119 mm.

Phyllosilicates

In phyllosilicates the [SiO$_4$] tetrahedra are linked to form infinite layers. In an infinite layer every [SiO$_4$] tetrahedron shares three of its four corners with other tetrahedra; the fourth corner (i.e. an O atom) is unshared. All unshared corners point in same direction, as shown in Figure 504. In the middle of the hexagon formed by the unshared corners there is room for an OH group. The Si/O ratio in the silicate layer is 2:5, which together with OH groups gives a formula ending with Si$_2$O$_5$(OH). The diversity in phyllosilicates is produced by various combinations of silicate layers, alone or in combination with layers of other types, as described below.

The layered structure of phyllosilicates is reflected in their physical properties. They normally occur as tabular or platy crystals; they have one pronounced cleavage direction parallel to the layers, and, because of the generally weak bonding between the layers, a relatively low hardness.

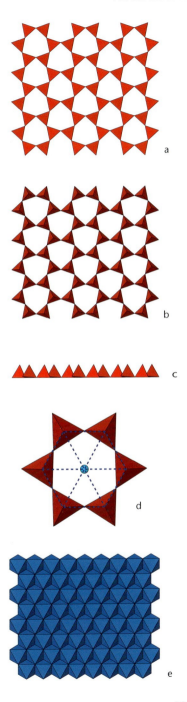

Figure 504. In phyllosilicates [SiO$_4$] tetrahedra are arranged in infinite layers in which every tetrahedron shares three corners with other tetrahedra, the fourth corner or O atom remaining unshared. In (a) the layer is viewed from one side, in (b) from the opposite side, and in (c) from the edge. Each layer consists of approximately hexagonal rings; in (d) an idealized ring is shown with an OH group inserted in the middle. The blue dotted lines indicate the triangles formed by connecting the O atoms and the central OH groups, most of which fit to the faces of the octahedra in the octahedral layer (e), where Mg or Al, for instance, are located.

331

Figure 505. In the crystal structure of antigorite a silicate layer (red) is connected to a layer of $[MgO_2(OH)_4]$ octahedra (blue). These composite layers are linked to corresponding layers by weak bonds. In the diagram the layers are shown as planar, but in the mineral they are more-or-less curved in order to achieve the best fit between the silicate and the octahedral layers. In chrysotile the layers are rolled up like a roll of wallpaper; this accounts for its fibrous character.

Figure 506. Chrysotile from Val Malenco, Lombardy, Italy. Subject: 94 × 148 mm.

Figure 507. Alternating layers of chrysotile and lizardite from Barberton, Mpumalanga, Republic of South Africa. Subject: 78 × 95 mm.

The serpentine group

The serpentine group includes three closely related minerals: *antigorite*, *lizardite*, and *chrysotile*. Basically they have the same crystal structure and chemical composition, but they differ from each other in the curvature of the layers, which results in antigorite and lizardite being dense or fine-grained and chrysotile being fibrous.

Antigorite, lizardite, and chrysotile, $Mg_3Si_2O_5(OH)_4$

Crystallography. Monoclinic, hexagonal, or orthorhombic, but macroscopic crystals do not occur; found only in dense masses (antigorite and lizardite) or fibrous (chrysotile).
Physical properties. Cleavage not detectable. Hardness 3–5; density about 2.6. Colour is usually greenish, often mottled, rarely yellow, brown, reddish, or greyish; greasy lustre, silky when fibrous; translucent.
Chemical properties, etc. Fe and Ni can to some extent replace Mg; Al can to a lesser extent replace Si or Mg. The basis of the crystal

Figure 508. Antigorite, pseudomorphous after olivine, from the Gardiner complex, Greenland. Subject: 71×76 mm.

Figure 509. Chrysotile var. mountain leather from Vallecas near Madrid, Spain. Subject: 103×116 mm.

Figure 510. Garnierite from Noumea, New Caledonia. Subject: 48 × 80 mm.

structure is a silicate layer connected to a layer of $[MgO_2(OH)_4]$ octahedra. This composite layer is linked to corresponding layers by weak bonds. Because of misfits between the points connecting the silicate layers and the $[MgO_2(OH)_4]$ octahedra, the composite layers may become bent in waves or rolled into fibres.

Names and varieties. *Garnierite* is an Ni-bearing serpentine with a characteristic green colour.

Occurrence. Serpentine minerals are widespread and occur especially as alteration products of olivine and other Mg-rich silicates. They are found in metamorphic as well as in igneous rocks and locally occur in considerable amounts.

Use. Massive forms of serpentine are used for ornamental work as building stones. For generations chrysotile has been the principal asbestos mineral because it is highly flexible, non-inflammable, and a poor conductor of heat. Large deposits of chrysotile are found e.g. in Thetford, Asbestos, and elsewhere in Quebec, Canada.

Diagnostic features. Mode of occurrence, the mottled colours, and greasy lustre. Chrysotile fibres are usually more flexible than fibres of amphibole asbestos.

The clay minerals

A large group of phyllosilicates are classified under the common name *clay minerals*. They are important rock-forming minerals, very fine-grained, and with the capacity to take up considerable amounts of H_2O and expand, which can cause the host rock to become plastic. They are characterized by wide chemical variation and they have, in varying degree, ion-exchanging properties. They are in general poorly crystallized and are commonly found in mixtures. For these and other reasons, clay minerals are difficult to identify without using special methods. Only four clay minerals are described here: *kaolinite* and, very briefly, *montmorillonite*, *illite*, and *vermiculite*.

Figure 511. Kaolinite, impure, from St. Stephens, Cornwall, England. Subject: 105 × 175 mm.

Kaolinite, $Al_2Si_2O_5(OH)_4$

Crystallography. Triclinic $\bar{1}$; macroscopic crystals not seen; mostly in loose or compact clays.
Physical properties. Cleavage perfect on {001}. Hardness about 2; density 2.6. Colour is white, also reddish or brownish owing to impurities; dull and earthy, but pearly lustre in coarser crystalline aggregates; translucent; feels greasy; becomes plastic in water.

Figure 512. Kaolinite, a pseudomorph after orthoclase, from St. Austell, Cornwall, England. Subject: 49 × 64 mm.

Figure 513. Vermiculite from Lenni, Pennsylvania, USA. Subject: 72 × 105 mm.

Chemical properties, etc. Kaolinite normally has a composition close to the ideal formula. It has a modified antigorite crystal structure: Mg^{2+} is in all three positions in the octahedral layer of antigorite, but in kaolinite Mg^{2+} is replaced by Al^{3+} in two of these positions, leaving the third empty.

Names and varieties. *Nacrite, dickite,* and *halloysite* are closely related minerals that are not distinguishable from kaolinite without special investigations.

Occurrence. Kaolinite is a widespread mineral. It constitutes the major part of kaolin and is an essential constituent of a series of other clays. It forms by weathering or hydrothermal alteration of Al-rich silicates, especially feldspars, and is in particular found in association with weathered gneisses and granites.

Use. Kaolinite is an important raw material and is used, e.g. as a filler in paper, for bricks and tiles, and (in pure forms) in porcelain.

Diagnostic features. Mode of occurrence, including impurities of quartz or partly weathered feldspar; greasy and plastic properties.

Montmorillonite,
$(Na,Ca)_{0,3}(Al,Mg)_2Si_4O_{10}(OH)_2 \cdot nH_2O$

The predominant clay mineral in bentonite, which is a clayey rock formed by the alteration of volcanic ash. Montmorillonite is of economic importance.

Illite, $(K,H_3O)Al_2(Si_3Al)O_{10}(H_2O,OH)_2$, a mineral of the mica group, is largely a hydrated and K-poor muscovite. It is the main constituent of most clayey sediments and shales.

Vermiculite,
$(Mg,Fe,Al)_3(Al,Si)_4O_{10}(OH)_2 \cdot 4H_2O$

Has a special ability to expand quickly and strongly on heating to nearly $300 \,°C$. It occurs primarily as an alteration product of phlogopite and biotite. Vermiculite is of economic importance.

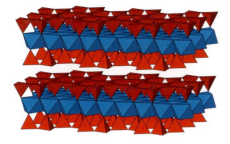

Figure 514. In talc, a layer of octahedra (blue) containing Mg is sandwiched between two layers of silicate (red). This composite layer is linked to corresponding layers by weak bonds.

The talc group

Talc, $Mg_3Si_4O_{10}(OH)_2$

Crystallography. Triclinic or monoclinic; crystals rare; mostly in platy or dense, fine-grained masses.

Physical properties. Cleavage perfect on {001}, cleavage folia flexible, but inelastic; dense masses very sectile. Hardness 1; density 2.8. Colour is usually pale green, also white or grey; greasy or pearly lustre, feels greasy; translucent.

Chemical properties, etc. Fe can replace Mg, but talc is normally rather pure. The crystal structure of talc consists of composite layers composed of octahedral layers containing Mg connected on both sides to silicate layers. These composite layers are linked by weak bonds, which explains the cleavage and hardness properties of talc.

Names and varieties. *Steatite* (an obsolete term) and *soapstone* generally refer a massive form of talc or to rocks almost entirely composed of it. In *minnesotaite* Fe predominates over Mg.

Occurrence. Talc is a characteristic mineral in low-grade metamorphic rocks rich in Mg, and locally talc can constitute nearly the whole rock. It occurs as a secondary mineral formed by alteration of ultrabasic rocks such as peridotites, dunites, and pyroxenites.

Use. Talc is a used as a filler in paint, rubber, paper, etc. Its poor thermal and electrical conductivity, non-inflammability, and resistance to acids make talc useful for many other industrial applications. One use is as a lubricant. It is familiar as talcum powder and is a popular material for carving.

Diagnostic features. The poor hardness and greasy character— but these are also properties of pyrophyllite.

Pyrophyllite, $Al_2Si_4O_{10}(OH)_2$

Crystallography. Triclinic or monoclinic; crystals rare, occasionally in typical radiating aggregates of flattened elongated crystals; mostly in dense talc-like masses.

Figure 515. Talc from New Hampshire, USA. Subject: 92×133 mm.

Figure 516. Pyrophyllite from Indian Gulch, California, USA. Field of view: 52 × 86 mm.

Physical properties. Cleavage perfect on {001}, cleavage folia slightly flexible, but inelastic; dense masses sectile. Hardness 1–1½; density 2.8. Colour is white, pale yellowish or greenish, sometimes more intensely coloured owing to impurities; greasy or pearly lustre, feels greasy; transparent to translucent.

Chemical properties, etc. The composition of pyrophyllite normally deviates only slightly from the ideal formula. It has a talc crystal structure, in which Al^{3+} replaces Mg^{2+} in two of the three positions in the octahedral layer of talc, leaving the third position empty.

Names and varieties. *Agalmatolite* is a dense variety used for carving, especially in China. The term *agalmatolite* can also refer to talc.

Occurrence. Pyrophyllite is less common than talc; it is found in Al-rich low-grade metamorphic rocks and is also known from hydrothermal quartz veins.

Use. Pyrophyllite and talc are used for similar purposes, but pyrophyllite is less important. It has been especially useful as a carrier in insecticides.

Diagnostic features. Pyrophyllite has most properties in common with talc; it is best recognized as crystals in radiating aggregates.

Mica group

Micas are important rock-forming minerals in igneous as well as in metamorphic rocks, and they are also widespread in many sediments. The crystal structure of mica can be described by taking a composite layer of talc or pyrophyllite as a starting point. By replacing one out of four Si^{4+} with Al^{3+} in the silicate layers, a deficit in charge balance is created, which can be compensated for by introducing a K^+ ion between the composite layers. In this way the composite layers are much more strongly linked together. Micas accordingly have a much greater hardness than talc or pyrophyllite. More details of the mica structure are given under muscovite and phlogopite.

Most micas are monoclinic, but because of the angles between prominent faces mica crystals usually have a pronounced hexagonal outline.

Micas are normally easily identified by their perfect cleavage on {001}, which can generate very thin cleavage folia that are flexible and are usually also elastic.

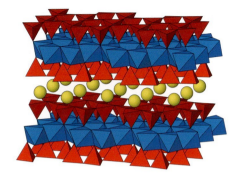

Muscovite, $KAl_2(Si_3Al)O_{10}(OH,F)_2$

Crystallography. Monoclinic $2/m$; crystals normally tabular parallel to {001} with uneven prism faces; sometimes conical habit; mostly in lamellar, platy, or scaly aggregates. **Physical properties.** Cleavage perfect on {001}, yielding very thin folia that are flexible and elastic. Hardness 2½; density 2.8. Colourless, rarely with pale yellow, green, or brown nuances; vitreous to pearly lustre; transparent.

Figure 517. The crystal structure of muscovite is built up of composite layers like those in pyrophyllite, in which a layer of octahedra (blue) is sandwiched between two silicate layers (red). In the octahedral layer Al is located in two-thirds of the available positions. leaving one-third empty. In the silicate layers of muscovite, one in four Si^{4+} is replaced by Al^{3+}, which opens up the possibility of accommodating K^+ ions (yellow) between the composite layers and linking them together relatively strongly. As the surfaces of the composite layers facing the K^+ are more-or-less hexagonal in symmetry, the two layers on either side of the K^+ ion can be orientated at angles of 0°, ± 60°, ± 120°, or 180° in relation to each other, thus resulting in the different polytypes.

Chemical properties, etc. Normally only minor substitutions take place in muscovite: mainly Na, Rb, or Cs replacing K, and Mg, Fe, Li, Cr, and V, etc. replacing Al. The crystal structure consists of composite layers of the pyrophyllite type, which is composed of an octahedral layer sandwiched between two silicate layers. In the octahedral layer Al^{3+} has entered two of three positions, leaving the third empty. In the silicate layers in muscovite one out of four Si^{4+} is replaced by Al^{3+}, which opens up the possibility of a K^+ being introduced between the composite layers and linking them tightly together.

Names and varieties. *Fuchsite* is a green Cr-bearing variety of muscovite.

Figure 518. Muscovite with quartz from Modum, Norway. Subject: 64 × 87 mm.

Figure 519. Phlogopite from Franklin, New Jersey, USA. Field of view: 30 × 45 mm.

Occurrence. Muscovite is a common rock-forming mineral. It occurs especially in granites and related pegmatites, in which it can reach considerable dimensions. It is also common in metamorphic rocks such as mica-schists and gneisses. Very fine-grained muscovite, *sericite*, is common as an alteration product of feldspars and other Al-rich silicates. As described under clay minerals, illite, which occurs as a major constituent of various clayey sediments, can be regarded as a hydrated, K-poor muscovite.

Use. Its low thermal capacity, non-inflammability, and good dielectric properties make muscovite useful in many applications, especially within the electrical industry. Powdered muscovite is used as a filler, e.g. in rubber and asphalt products.

Diagnostic features. The perfect cleavage, flexible and elastic cleavage folia, and light colour are diagnostic. Darker-coloured muscovite is difficult to distinguish from phlogopite.

Paragonite, $NaAl_2(Si_3Al)O_{10}(OH)_2$

Paragonite has Na^+ between the composite layers instead of K^+ as in muscovite. It occurs mainly in white, sometimes pale yellow or green, fine-grained scaly aggregates with pearly lustre. It is the predominant mineral in paragonite-schists, which are the host rocks for several occurrences of staurolite and kyanite.

Glauconite, $(K,Na)(Fe,Al,Mg)_2(Si,Al)_4O_{10}(OH)_2$

Glauconite is a mica structurally related to muscovite but showing far more variation in chemistry. It occurs mostly as small green or greenish-black grains in loosely consolidated marine sediments such as greensands. *Celadonite* is a closely related mineral; it occurs primarily as green earthy cavity fillings in basalts, e.g. in the Faroe Islands.

Phlogopite, $K(Mg,Fe)_3(Si_3Al)O_{10}(F,OH)_2$

Crystallography. Monoclinic $2/m$; crystals usually tabular parallel to {001} or in pseudo-hexagonal prisms, often conical; mostly as lamellar or platy aggregates.

Physical properties. Cleavage perfect on {001}, can be cleaved into very thin folia that are flexible and elastic. Hardness 2½–3; density 2.9, increasing with Fe content. Colour is pale yellow to brown, rarely greenish or reddish; vitreous to pearly lustre; transparent in thin folia.

Chemical properties, etc. A complete solid-solution series exists between phlogopite and *annite*, $KFe_3(Si_3Al)O_{10}(OH,F)_2$, in which biotite in practice designates the Fe-rich part of the series. Small amounts of Na, Ca, Rb, or Cs can replace K, whereas Fe^{2+} is the principal substitute for Mg. The crystal structure of phlogopite corresponds to that of muscovite except for the occupation of all octahedral positions as opposed to only two-thirds in muscovite.

Occurrence. Phlogopite is found in Mg-rich rocks such as peridotites and other ultrabasic rocks and in metamorphosed dolomites.

Use. Phlogopite has almost the same properties as muscovite and therefore largely the same applications.

Diagnostic features. The perfect cleavage with flexible and elastic cleavage folia readily identify phlogopite as a mica. Colour and mode of occurrence are the best clues in distinguishing muscovite, phlogopite, and biotite.

Biotite, $K(Mg,Fe)_3(Si_3Al)O_{10}(OH,F)_2$

Crystallography. Monoclinic $2/m$; good crystals uncommon, usually tabular parallel to {001} or short prismatic; mostly in irregular platy or scaly aggregates.

Physical properties. Cleavage perfect on {001}, cleavage folia flexible and elastic. Hardness 2½–3; density from about 2.8 upwards, increasing with Fe-content. Colour is

Figure 521. Biotite from Arendal, Norway. Field of view: 59×85 mm.

dark brown, green, or black; vitreous to submetallic lustre; transparent to translucent.

Chemical properties, etc. A complete solid-solution series exists between phlogopite and *annite*, $KFe_3(Si_3Al)O_{10}(OH,F)_2$, in which biotite in practice represents the Fe-rich part of the series. Small amounts of Na, Ca, Ba, Rb, or Cs can replace K, whereas Fe^{2+} and to a lesser extent Fe^{3+}, Al, Mn, and Ti substitute for Mg. The crystal structure is as for phlogopite.

Names and varieties. *Lepidomelane* is a particularly Fe-rich and deep black biotite.

Occurrence. Biotite is widespread in many types of rocks. It is found in igneous rocks ranging from Si-rich rocks, such as granites, to more basic rocks, such as gabbros. It is also an important constituent in many meta-

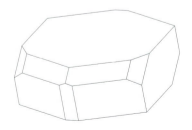

Figure 520. Biotite: {010}, {001}, {10$\bar{1}$}, {11$\bar{1}$}, {112}, and {132}.

Figure 522. Lepidolite from Tørdal, Telemark, Norway. Subject: 83 × 105 mm.

morphic rocks, such as mica-schists, gneisses, and hornfels.

Diagnostic features. The perfect cleavage, flexible and elastic cleavage folia, and the (usually) dark colour.

Lepidolite, $K(Li,Al)_3(Si,Al)_4O_{10}(F,OH)_2$

Crystallography. Monoclinic $2/m$; crystals uncommon, normally in small plates with hexagonal outline; mostly as scales in more-or-less fine-grained aggregates.

Physical properties. Cleavage perfect on {001}, cleavage folia flexible and elastic. Hardness about 3; density about 2.8. Colour is lilac or pink, rarely colourless, grey, or yellowish; pearly lustre; transparent to translucent.

Chemical properties, etc. Na, Rb, or Cs can to some extent replace K, whereas the ratio between Al and Li in the octahedral layer varies somewhat. Lepidolite has a muscovite crystal structure.

Names and varieties. *Polylithionite*, $KLi_2Al-Si_4O_{10}(F,OH)_2$, is closely related to lepidolite, but has a higher content of Li and Si. In addition to the common properties of a mica, polylithionite is characterized by a light greenish colour. In some instances polylithionite is developed as hexagonal tabular crystals divided into six sectors, sometimes in intergrowths with epistolite.

Occurrence. Lepidolite is a characteristic mineral in granite pegmatites, often in association with other Li-bearing minerals such as tourmaline or spodumene. Lepidolite is known from, e.g. Auburn, Maine, and from Little Three mine and other mines in San Diego County, California, USA. Polylithionite is a typical mineral of some nepheline-syenite pegmatites, and is known from, e.g. the Ilímaussaq complex in Greenland.

Use. Lepidolite has been mined as a source of Li.

Diagnostic features. Mode of occurrence and colour, together with the common properties of micas.

Zinnwaldite,
$K(Al,Fe,Li)_3(Si,Al)_4O_{10}(OH)F$

Like polylithionite, related to lepidolite, but contains more Fe at the expense of Li. Well-developed crystals are more often seen in zinnwaldite than in lepidolite, most frequently as tabular crystals. The physical properties are as for lepidolite, but the colour is more often silver-grey, yellowish, brown, or dark green to nearly black. Zinnwaldite is primarily known from pneumatolytic veins in granites, where it typically occurs in association with cassiterite.

Figure 523. Polylithionite from Kangerdluarsuk, Ilímaussaq complex, Greenland. Subject: 83 × 113 mm.

343

Margarite, $CaAl_2(Si_2Al_2)O_{10}(OH)_2$

Crystallography. Monoclinic *m*; crystals rare, mostly in platy or scaly aggregates.

Physical properties. Cleavage perfect on {001}, cleavage folia are brittle in contrast to the common micas. Hardness about 4; density 3.0. The colour is white, reddish, or pearly-grey; pearly lustre; translucent.

Chemical properties, etc. Margarite is primarily characterized by having Ca between the composite layers, in contrast to most other micas that have K+. The divalent Ca^{2+} link the layers more strongly than the monovalent K+, and this is reflected in the greater hardness of margarite. The charge balance is maintained by the replacement of two rather than one Si^{4+} by Al^{3+}.

Names and varieties. Margarite belong to a group called the *brittle micas*, which includes micas with brittle cleavage folia.

Occurrence. Margarite is a relatively rare mineral primarily known from emery deposits, in which it is associated with corundum and diaspore, as at Chester, Massachusetts, USA.

Diagnostic features. Mode of occurrence and, in relation to other micas, the brittle cleavage folia and greater hardness.

Clintonite, $Ca(Mg,Al)_3(Al,Si)_4O_{10}(OH,F)_2$

Clintonite belongs to the brittle micas. In the same way that margarite can be considered as a Ca analogue to muscovite, clintonite is close to being a Ca analogue to phlogopite. Clintonite, with properties closely resembling those of margarite, can be reddish, yellow, or deep green. It occurs sporadically in some limestones, talc, and chlorite-schists.

Figure 524. Margarite from Chester, Massachusetts, USA. Field of view: 58 × 87 mm.

Figure 525. Clintonite from Nicolau-Maxilianowicz mine, Akhmatovsk, Urals, Russia. Subject: 67×73 mm.

The chlorite group

The chlorite group of closely related phyllosilicates is widespread. They occur primarily in weakly metamorphosed rocks, as alteration products of pyroxenes, amphiboles, and micas in igneous rocks, and as important constituents of many sediments. Despite considerable chemical variation within the group, the physical properties of its members are rather uniform. This implies that in practice these minerals are extremely difficult to distinguish from each other. For this reason, only one chlorite mineral is here described in detail.

Clinochlore, $(Mg,Fe)_5Al(Si_3Al)O_{10}(OH)_8$

Crystallography. Triclinic or monoclinic; good crystals rare, mostly in pseudo-hexagonal tabular crystals with predominant {001}; mostly in platy or fine-scaly aggregates or dispersed as grains.

Physical properties. Cleavage perfect on {001}, cleavage folia flexible, but inelastic. Hardness about 2½; density normally 2.7–2.9, higher with increasing Fe content. Colour is usually green (hence the name chlorite), rarely yellowish, brown, or violet; vitreous lustre, often dull; translucent.

Chemical properties, etc. The crystal structure of clinochlore consists of a brucite-like layer, i.e. a layer of $[Mg_2Al(OH)_6]$ octahedra,

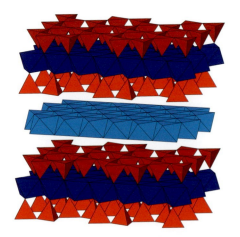

Figure 527. Chlorite: {001}, {110}, {10$\overline{1}$}, {130}, {041}, and {11$\overline{1}$}.

Figure 526. The crystal structure of clinochlore is built up of layers of talc, i.e. composite layers of two silicate layers (red) and an octahedral layer (dark blue), with layers of [Mg$_2$Al(OH)$_6$] octahedra (lighter blue) inserted between them. The inserted layers are identical in composition and structure to the mineral brucite. To emphasize this arrangement of layers, the formula of clinochlore could be rewritten as (Mg,Al)$_3$(OH)$_6$·(Mg,Al)$_3$(Si,Al)$_4$O$_{10}$(OH)$_2$, the part of the formula before the dot being the inserted layer and the part after the dot the talc layer.

sandwiched between two talc-type layers. In order to elucidate this arrangement the formula given above could be rewritten as (Mg,Al)$_3$(OH)$_6$·(Mg,Al)$_3$(Si,Al)$_4$O$_{10}$(OH)$_2$, in which the central brucite-like layer stands before the dot and the talc-layer part after it. In total there are six (Mg,Al) positions, in which the Mg/Al ratio can vary and in which Fe^{2+}, Fe^{3+}, and several other less important elements can replace (Mg,Al). As Al can also to a variable degree replace Si, it is apparent that substantial chemical variation among chlorite minerals is possible.

Names and varieties. *Chamosite*, an Fe analogue to clinochlore, and *thuringite*, also an Fe-rich chlorite, are important constituents of some sedimentary iron ores that have been of some economical importance. *Sudoite* is a chlorite particularly rich in Al. *Cookeite* is

Figure 528. Chlorite from Modum, Norway. Subject: 59 × 74 mm.

Figure 529. Clinochlore from Val di Vizze (formerly Pfitschtal, Austria), Italy. Subject: 63 × 97 mm.

Figure 531. Cookeite with quartz from Saline County, Arkansas, USA. Subject: 62 × 91 mm.

a rare Li-bearing chlorite. *Kämmererite* is a Cr-rich clinochlore with a characteristic purple colour.

Occurrence. Chlorite minerals are important constituents of low-grade metamorphic rocks and are indicative of the greenschist facies; they are predominant in chlorite-schists. They are also common as alteration products of biotite and other silicates containing Fe and Mg. The green colour so common in many rocks is often caused by chlorite formed by disintegration of the primary minerals. Chlorite is also common in many sediments.

Diagnostic features. Colour, cleavage, and the non-elastic cleavage folia.

Figure 530. Kämmererite from Anatolia, Turkey. Field of view: 32 × 48 mm.

347

Figure 532. Apophyllite from Nasik, India. Subject: 35 × 62 mm.

Apophyllite, $KCa_4Si_8O_{20}(F,OH)\cdot8H_2O$

Crystallography. Tetragonal $4/m2/m2/m$; well-developed crystals common, mostly in combinations of {110} and {011}, sometimes also {001}; some crystals dominated by {110} and {001} have a pseudo-cubic habit; prism commonly striated parallel to the c axis.

Physical properties. Cleavage perfect on {001}. Hardness 5; density 2.4. Colourless, white, or grey, rarely with pale green or yellow nuances; vitreous lustre, pearly on {001}; transparent to translucent.

Chemical properties, etc. F can predominate over OH and vice versa, which in principle gives rise two distinct minerals: *fluorapophyllite* and *hydroxyapophyllite*. Apophyllite

is a phyllosilicate in which the silicate layers consist of combined fourfold and eightfold rings of $[SiO_4]$ tetrahedra.

Occurrence. Apophyllite is a secondary mineral occurring in fissures and cavities in basalts and similar rocks in association with zeolites, calcite, etc. Excellent crystals are found in Iceland and in particular in Poona, Nasik, and other districts in India.

Diagnostic features. Crystal habit, mode of occurrence, the difference in lustre of {001} and other faces, and sometimes colour.

Prehnite, $Ca_2Al(Si,Al)_4O_{10}(OH)_2$

Crystallography. Orthorhombic $2mm$; single crystals rare, mostly tabular parallel to {001}; usually stalactitic or botryoidal with crests of small crystals; also massive.

Physical properties. Cleavage distinct on {001}; Hardness 6–6½; density 2.9. Colour is light green to darker green, also white or grey; vitreous lustre; translucent.

Chemical properties, etc. Small amounts of Fe^{3+} can replace Al.

Occurrence. Prehnite is a secondary mineral

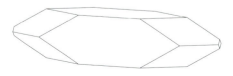

Figure 533. Apophyllite: {110}, {011}, and {001}; the c axis is almost horizontal.

found particularly in cavities and fissures in basalts and similar rocks. It typically occurs in association with zeolites, calcite, or pectolite. Fine crystals are found at Jeffrey mine, Asbestos, Quebec, Canada.

Diagnostic features. Habit, colour, and mode of occurrence.

Pyrosmalite, $(Fe,Mn)_8Si_6O_{15}(OH,Cl)_{10}$

Trigonal; occurs as tabular or columnar crystals; also in dense or fine-grained masses. Cleavage perfect on {0001}; hardness 4–4½, density about 3.1. Colour is brownish- to olive-green; greasy to submetallic lustre; translucent. Pyrosmalite is found in association with Fe- or Mn-rich ore deposits and is known, e.g. from Nordmark in Sweden and from Broken Hill in New South Wales, Australia.

Figure 535.
Pyrosmalite from Nordmark,
Värmland, Sweden. Subject: 49 × 78 mm.

Figure 534. Prehnite from Brandenberg, Namibia. Field of view: 90 × 135 mm.

Petalite, $LiAlSi_4O_{10}$

Monoclinic; crystals rare, mostly in large feldspar-like aggregates. Cleavage perfect on {001}; hardness 6½; density about 2.4. Colourless, white, or grey, rarely reddish or greenish; vitreous lustre; transparent to translucent. Petalite is found in granite pegmatites and is known, e.g. from Utö and Varuträsk, Sweden.

Gyrolite, $NaCa_{16}AlSi_{24}O_{60}(OH)_8 \cdot 14H_2O$

Triclinic; forms small tabular pseudo-hexagonal crystals in small lamellar or spherical aggregates. Cleavage perfect on {001}, cleavage folia brittle; hardness 3–4; density 2.4. Colourless, white, or greyish with greenish or brownish tinge; pearly lustre; transparent to translucent. *Reyerite* is closely related to gyrolite. Both minerals occur in cavities in bas-alts, and are known, e.g. from Niaqornat in Greenland and in fine crystals from Poona, India.

Stilpnomelane, $(K,Ca,Na)(Fe,Mg,Al)_8(Si,Al)_{12}(O,OH)_{36} \cdot nH_2O$

Triclinic; good crystals rare, mostly as mica-like plates in bundles. Cleavage perfect on {001}, cleavage folia brittle; hardness 3–4; density about 2.9. Colour is dark green to black, also golden- or reddish-brown; strong vitreous lustre, almost submetallic; mostly translucent. The composition varies: e.g. Mn can replace Fe. Stilpnomelane occurs in Fe-rich regional metamorphic rocks associated with chlorite and epidote, and in metamorphic Fe-ore deposits, typically with other Fe-rich silicates. It is known, e.g. from Stirling mine, Antwerp, New York State, USA.

Figure 536. Gyrolite from Qaarusuit, Greenland. Subject: 37 × 62 mm.

Figure 537. Stilpnomelane from Switzerland. Field of view: 37 × 72 mm.

Chrysocolla,
(Cu,Al)$_2$H$_2$Si$_2$O$_5$(OH)$_4$·nH$_2$O

Presumably orthorhombic; mostly in dense cryptocrystalline incrustations, also botryoidal, stalactitic, fibrous, earthy. No cleav-age; hardness 2–4; density 1.9–2.4. Colour is green or blue in various shades, also brown or black owing to impurities; streak bluish-white; lustre mostly wax-like; almost opaque. Chrysocolla is found in the oxidized zones of Cu deposits, commonly in association with

Figure 538. Chrysocolla from Tintic District, Utah, USA. Subject: 64 × 113 mm.

351

Figure 539. Palygorskite from Lemesurier Island, Alaska, USA. Subject: 67×133 mm.

malachite and azurite. It sometimes resembles malachite in habit and colour but does not effervesce with hydrochloric acid.

Palygorskite,
$(Mg,Al)_2Si_4O_{10}(OH)\cdot 4H_2O$

Monoclinic or orthorhombic; occurs in tiny ruler-shaped crystals, less than 0.3 mm long, assembled in leather-like aggregates, commonly very porous. Hardness 2–2½; density 2.3. Colour is white, grey, yellowish-brown, or greenish; dull lustre. Palygorskite is found in hydrothermal veins in basalts, granites, and syenites. It is sometimes confused with other minerals forming 'mountain-leather', e.g. chrysotile.

Naujakasite, $Na_6(Fe,Mn)Al_4Si_8O_{26}$

Monoclinic; occurs as mica-like crystals with a rhombic outline. It has one perfect cleavage direction, hardness 2-3, and density 2.6. The colour is light greenish- or silver-white; pearly lustre. Naujakasite is known from Naujakasik in the Ilímaussaq complex, Greenland.

Cavansite, $Ca(VO)Si_4O_{10}\cdot 4H_2O$

Orthorhombic; found as blue to bluish-green prismatic or tabular crystals in rosettes. Hardness 3–4; density 2.3. It occurs in cavities in basalts and tuffs, usually in association with zeolites and is known *e.g.* from Poona, India.

Sepiolite, $Mg_4Si_6O_{15}(OH)_2\cdot 6H_2O$

Orthorhombic; like palygorskite forms tiny ruler-shaped or fibrous crystals, but in sepiolite these are assembled in strongly porous nodules. The mineral itself has a hardness of 2–2½, but the porous aggregates are softer; likewise, the density of the mineral is about

Figure 540. Naujaka-site from Naujakasik, Ilímaussaq complex, Greenland. Field of view: 38 × 38 mm.

Figure 541. Cavansite from Poona, India. Field of view: 33 × 58 mm.

2.1, much more than the porous aggregates that float on water. Sepiolite nodules stick firmly to the tongue. The colour is white, grey, or light yellowish; dull lustre; almost opaque. *Meerschaum* is an old name for sepiolite, as used in ornamental carvings, tobacco pipes, etc. Sepiolite occurs as an alteration product of serpentine; serpentine breccias at Eskişehir, Anatolia, Turkey, are a noted locality.

Figure 542. Sepiolite from Boskovice, Moravia, Czech Republic. Subject: 88 × 98 mm.

Tectosilicates

In tectosilicates all four oxygen atoms of the $[SiO_4]$ tetrahedra are shared with other $[SiO_4]$ tetrahedra in a three-dimensional lattice. As every O atom is bonded to two Si atoms, the Si/O ratio becomes 1:2 as in quartz, SiO_2. The diversity in other tectosilicates is created by the partial substitution of Al^{3+} for Si^{4+}, by up to 50%. In this way a deficit in positive charge is established, which can be balanced by the introduction of large cations such as K^+, Na^+, or Ca^{2+}. These ions are accommodated in cavities in the relatively open three-dimensional lattice as in the feldspars albite, $NaAlSi_3O_8$, and anorthite, $CaAl_2Si_2O_8$.

The tectosilicates, which include several of the principal rock-forming minerals, constitute a relatively homogeneous group of minerals. They are typically colourless or only pale coloured, and have a vitreous lustre and a relatively low density as a result of the open structure; their hardness is generally between 5 and 7, but is lower for some zeolites.

The SiO$_2$ group

This group includes *quartz, tridymite, cristobalite, stishovite,* and *coesite,* all with the formula SiO_2. Quartz is widespread in many types of rocks, whereas tridymite and cristobalite are common only in Si-rich volcanic rocks. Stishovite and coesite are rare minerals found at the sites of meteorite impacts. The group also includes the amorphous and water-containing mineral *opal.*

Figure 543. Mesolite from Palmerston, North Otago, South Island, New Zealand. Field of view: 24 × 36 mm.

Quartz, SiO₂

Crystallography. Trigonal 32; crystals common, commonly prismatic, consisting of a hexagonal prism, $\{10\bar{1}0\}$, terminated by two rhombohedra, $\{10\bar{1}1\}$ and $\{01\bar{1}1\}$, together resembling a hexagonal bipyramid when uniformly developed; to this can be added a trigonal bipyramid, $\{2\bar{1}\bar{1}1\}$, and occasionally the general crystal form, a trigonal trapezohedron, $\{hkil\}$. The trapezohedron reveals the true symmetry of quartz and shows whether the crystal is right-handed or left-handed; when a trapezohedral face is located above and to the right in relation to a prism face viewed from the front, the crystal is right-handed; in the opposite case, it is left-handed. Prism faces generally horizontally striated. Twinning is very common but not always visible; common twin laws are *Dauphiné* with the *c* axis as twin axis, and *Brazil* with $\{11\bar{2}0\}$ as twin plane. Quartz is very common in fine- to coarse-grained aggregates and in microcrystalline forms described below.

Physical properties. No cleavage, fracture conchoidal. Hardness 7; density 2.65, slightly lower in microcrystalline varieties. Usually colourless or white, but all colours occur owing to impurities; vitreous lustre, slightly greasy in microcrystalline varieties; transparent to translucent. Strongly piezoelectric.

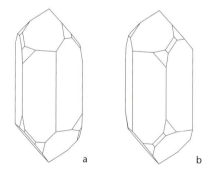

Figure 544. Quartz: (a) left-handed quartz with $\{10\bar{1}0\}$, $\{10\bar{1}1\}$, $\{01\bar{1}1\}$, $\{2\bar{1}\bar{1}1\}$, and $\{6\bar{1}\bar{5}1\}$; (b) right-handed quartz with $\{10\bar{1}0\}$, $\{10\bar{1}1\}$, $\{01\bar{1}1\}$, $\{11\bar{2}1\}$, and $\{51\bar{6}1\}$.

Figure 545. Quartz: (a) twinned on the Brazil law with $\{11\bar{2}0\}$ as twin plane; (b) twinned on the Dauphiné law with the *c* axis as twin axis.

Chemical properties, etc. Quartz is always fairly pure; the impurities causing colour are present in only very small amounts. In the crystal structure of quartz the $[SiO_4]$ tetrahedra are arranged in helices parallel with the *c* axis. The helices are either right- or left-handed and are linked sideways so that every $[SiO_4]$ tetrahedron shares all four corners with others.

Names and varieties. Quartz occurs in a large number of varieties. They are divided into common macrocrystalline varieties, in which the colour is the essential criterion, and microcrystalline varieties, consisting of small fibrous or granular crystallites. The macrocrystalline varieties include the following. *Rock crystal*, colourless and completely transparent. *Milky quartz*, the common white or grey quartz known from pegmatites. *Amethyst*, which displays shades of violet owing to the presence of small amounts of Fe^{3+}; it is a popular gemstone. *Citrine*, a yellow quartz resembling topaz. *Smoky quartz*, smoky-yellow or brownish, also resembling topaz. *Rose quartz*, rose-red or pink; seldom found as good crystals. *Cat's eye*, quartz with inclusions of asbestos or other fibres (but also used for other minerals). *Aventurine* has inclusions of mica or hematite. The name 'aventurine' is sometimes also used for a feldspar with similar inclusions. *Chalcedony* is partly a collective name for the microcrystalline varieties, and partly the name of the ordinary grey or brownish, wax-like and translucent variety so common in stalactitic or botryoidal forms. Among the especially coloured varieties are *carnelian*, which is red, and *sard*, which is browner. *Chrysoprase* is green in various shades, *plasma* is dark green, and *heliotrope* green with red spots. *Agate* is banded chalcedony, in which differently coloured layers alternate, usually in concentric forms. In *moss agate* the colour, usually brown or dark green, varies owing to the presence of dendritic impurities, typically Mn oxides. *Onyx* is banded like agate but in plane parallel layers; in *sardonyx* white layers alternate with brown or black layers. In both agate and onyx layers of opal can be present. *Silicified* or *petrified wood* normally consists of chalcedony. The granular microcrystalline varieties include *flint*, which is found as grey or black concretions in limestones, especially chalk; *chert*, lighter-coloured, and *jasper*, which is mostly red, brown, or green, sometimes banded. *Tiger's eye* is a variety of quartz in which an asbestos mineral has been replaced by quartz while preserving the fibrous structure, resulting in a special silky lustre.

Occurrence. Quartz is widespread in many geological environments. It is an essential mineral in many metamorphic rocks such as gneisses, mica-schists, quartzites, and eclogites, and in Si-rich igneous rocks such as granites and granodiorites, including the related pegmatites and veins. Owing to its chemical resistance and great hardness, quartz is the principal mineral in many unconsolidated or consolidated sediments, such as sandstones, conglomerates, etc.

Use. Quartz has many uses. In the construction industries it is used in concrete, cement, mortar, and, in the form of sandstone, as a building stone. It is used in the production of glass, porcelain, and similar materials, as an abrasive, a filler, etc. Quartz has special properties that are useful in the optical industry, and its piezoelectric properties are used for controlling frequencies, e.g. in watches. Natural quartz is commonly twinned and is thus unsuitable for these purposes, so such material is produced synthetically. Finally, quartz and its many varieties are popular as gemstones and for decoration.

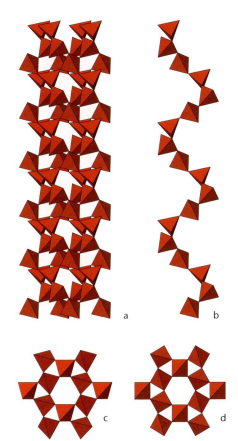

Figure 546. In the crystal structure of quartz the [SiO$_4$] tetrahedra are arranged in helices along the *c* axis. The helices are right- or left-handed and are linked so that every [SiO$_4$] tetrahedron shares all four corners with others; (a) shows a section of the structure with the *c* axis vertical; (b) shows one of the helices from (a) to give a better view of the helix; (c) the section from (a) viewed along the *c* axis, demonstrating the trigonal symmetry; (d) a corresponding section of high quartz viewed along the *c* axis. High quartz is the stable polymorph above 573 °C. It is hexagonal; the helices are slightly adjusted as compared with those of low quartz, which is the stable form at less than 573 °C. The transition from one polymorph to the other is reversible and takes place almost momentarily by small adjustments without any breaking of bonds.

Diagnostic features. Crystal habit, conchoidal fracture, and hardness.

Figure 547. Quartz
var. rock crystal from
Brazil. Field of view:
60 × 85 mm.

Figure 548. Quartz var.
amethyst from Hanekleiv-
tunnelen, Holmestrand,
Vestfold, Norway.
Subject: 42 × 50 mm.

Figure 549. Quartz var. rock crystal from Kongsberg, Norway. Field of view: 32 × 48 mm.

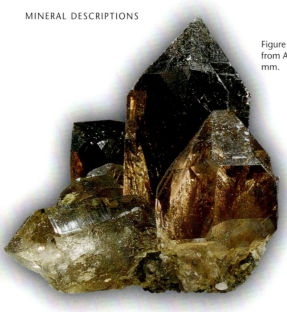

Figure 550. Quartz var. smoky quartz from Alençon, France. Subject: 53 × 53 mm.

Figure 551. Quartz var. rose quartz from Rabenstein, Bayern, Germany. Subject: 74 × 105 mm.

Figure 552. Quartz var. citrine from Schwarzenstein, Tirol, Austria. Field of view: 18 × 30 mm.

Figure 555. Quartz var. chalcedony var. flint (black) from Stevns Klint, Denmark. Field of view: 50 × 67 mm.

Figure 553. Quartz var. chalcedony from the Faeroe Islands. Subject: 92 × 122 mm.

Figure 554. Quartz var. chalcedony var. chrysoprase from Szklary, Poland. Field of view: 70 × 120 mm.

Figure 556. Quartz var. chalcedony var. jasper from Urals, Russia. Subject: 77 × 156 mm.

Figure 557. Quartz var. chalcedony var. carnelian from Ballast Point, Tampa, Florida, USA. Subject: 34 × 41 mm.

Figure 558. Quartz var. chalcedony var. agate from an unknown locality. Subject: 72 × 108 mm.

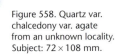

Figure 559. Cristobalite in obsidian from Coso Hot Springs, Inyo County, California, USA. Subject: 66×115 mm.

Tridymite, SiO_2

Orthorhombic, found as small (≤ 1 mm) pseudo-hexagonal tabular crystals, often twinned. Hardness 7, density 2.3, i.e. lower than quartz; appearance as quartz. Tridymite is common in some Si-rich volcanic rocks, such as rhyolites and obsidians, usually in association with sanidine and cristobalite. The mode of occurrence and crystal habit can be characteristic, but tridymite is not otherwise easily identified without special investigations.

Cristobalite, SiO_2

Tetragonal and found as small crystals with octahedron-like habit, which is preserved from the cubic high-temperature cristobalite; crystals commonly assembled in spherical aggregates. The physical properties are almost the same as for tridymite. Cristobalite commonly occurs in Si-rich volcanic rocks like obsidians, where it is present in the fine-grained matrix as well as in cavities forming small crystal groups.

Opal, $SiO_2 \cdot nH_2O$

Crystallography. Amorphous; occurs massive in botryoidal, stalactitic, or encrusted forms.

Figure 560. Layers of opal alternating with layers of chalcedony, the Faeroe Islands. Field of view: 72×88 mm.

Physical properties. Conchoidal fracture. Hardness 5–6; density 2.0–2.2, according to water content. Colourless, white, or grey; rarely in yellowish, brownish, reddish, greenish, or bluish shades. Opal commonly has a milky tinge (opalescence); precious opal has a particular play of colour (iridescence); vitreous lustre, sometimes dull, wax-like; translucent to transparent.

Chemical properties, etc. The water content in opal varies; it is typically 3–9 wt % but can be up to about 20 wt %. Even though opal is not crystalline in the usual meaning of the word, order of some kind exists: opal consists of a dense packing of spheres, typically 1500–3000 Å in diameter. In common opal the size of the spheres varies greatly, whereas the spheres in precious opals are so uniform in size that they form a regular lattice, in which the light is refracted. In some opals the spheres themselves are completely amorphous, whereas in other opals the spheres are partially crystalline and consist of disordered layers of tridymite and cristobalite.

Names and varieties. *Common opal* does not have a play of colour, as in *precious opal*. Precious opal includes *black opal* with a black ground colour, and *fire opal* with an intense play of red or orange colours. *Hyalite* is a completely transparent and colourless opal. *Silicified* or *petrified wood* can consist of opal but is more often chalcedony.

Occurrence. Opal occurs in fissures and cavities in various rocks, where it is deposited by hydrous solutions at relatively low temperatures. It occurs as siliceous sinter or 'geyserite' at hot springs and geysers and constitutes the majority of diatomites, which are fine-grained chalk-like deposits formed by the accumulation of shells from diatoms.

Use. Precious opal is an appreciated gemstone; most of the gems derive from sandstone deposits in Australia. Diatomite is used as an insulation material, an abrasive, and in filtration.

Diagnostic features. Opal has slightly lower hardness and density than chalcedony, which it resembles in other respects.

Figure 561. Opal var. silicified wood from Clover Creek, Idaho, USA. Subject: 147 × 202 mm.

Figure 562. Opal var. precious opal from Queensland, Australia. Field of view: 40 × 60 mm.

The feldspar group

The feldspars are the most widespread group of minerals; they alone constitute more than half the earth's crust. They are essential minerals in most igneous and metamorphic rocks, and are also abundant in sediments.

Feldspars are divided into two subgroups: (1) *potassium* or *K-feldspars*, including *sanidine*, *orthoclase*, and *microcline*, all $KAlSi_3O_8$; (2) *plagioclases*, consisting of a solid-solution series between two end-members, *albite*, $NaAlSi_3O_8$, and *anorthite*, $CaAl_2Si_2O_8$. As a solid-solution series exists between K- and Na- feldspars, they are frequently bracketed together as alkali feldspars. In addition, there are the rather rare Ba feldspars.

All feldspars basically have the same crystal structure: a three-dimensional framework of $[(Si,Al)O_4]$ tetrahedra, in which there are spaces large enough to accommodate K, Na, Ca, or Ba. In the K-feldspars this structure is monoclinic or nearly monoclinic, whereas in plagioclases it is triclinic. Further structural details are given under sanidine and albite.

Feldspars are rather uniform in their physical properties. They all have two good cleavage directions perpendicular or almost perpendicular to each other, a hardness of about 6, and a density of about 2.6.

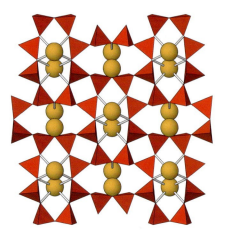

Figure 563. In sanidine the crystal structure consists of a three-dimensional network of [(Si,Al)O$_4$] tetrahedra (red) with cavities containing the large K ions (orange). Si and Al are randomly distributed. The drawing shows a slice of the structure seen along the *a* axis, with the *b* axis E–W. Because only a slice is shown here, the nine bonds between K and O (the tetrahedral corners) cannot be seen, nor is it evident that all corners of the tetrahedra are shared with other tetrahedra. (Compare with Figure 575.)

Sanidine, KAlSi$_3$O$_8$

Crystallography. Monoclinic 2/*m*; crystals commonly tabular parallel to {010}, crystal habit and twinning as for orthoclase (see below).

Physical properties. Cleavage perfect on {001}, good on {010}. Hardness 6; density 2.6. Colourless, white, or grey; can have a bluish milky tinge in some directions (iridescence); vitreous lustre, pearly on cleavage surfaces; often transparent.

Chemical properties, etc. At high temperatures there is a complete solid-solution series between sanidine and albite, NaAlSi$_3$O$_8$. The part of this series with Na > K is triclinic and is called *anorthoclase*. The crystal structure of sanidine consists of a three-dimensional framework of [(Si,Al)O$_4$] tetrahedra with spaces containing the large K ions. The structure is monoclinic and differs from that of

the triclinic feldspars, partly in being less distorted and 'collapsed', as in albite, and partly by having Si and Al ions in a completely disordered state, i.e. they are randomly distributed in the tetrahedral positions. This structure is the stable polymorph above about 700 °C. Sanidine formed in volcanic lavas is preserved because of the rapid cooling of these rocks during their formation.

Names and varieties. *Moonstone* is an iridescent feldspar, e.g. sanidine; it is used as a gemstone.

Occurrence. Sanidine is a characteristic mineral in rhyolites and similar extrusive igneous rocks rich in K and Si that were rapidly cooled when they were formed. It is commonly present in the rock as distinct crystals in a fine-grained matrix; a classic example is known from Drachenfels in Germany.

Diagnostic features. Crystal habit, cleavage, and mode of occurrence; sanidine and orthoclase do not conform to the twin laws for triclinic feldspars.

Figure 564. Sanidine in trachyte from Drachenfels, Germany. Subject: 71 × 87 mm.

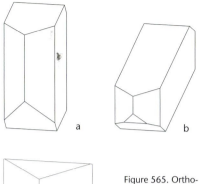

Figure 565. Orthoclase: (a) {010}, {001}, {110}, and {20$\bar{1}$}; (b) {010}, {001}, {110}, {10$\bar{1}$}, and {20$\bar{1}$}; (c) var. adularia: {001}, {110}, and {10$\bar{1}$}.

a special play of colour with strong bluish and yellowish colours (labradorescence); vitreous lustre, pearly lustre on cleavage surfaces; mostly translucent.

Chemical properties, etc. Orthoclase usually contains some Na. At high temperatures there is miscibility between the two end-members, orthoclase, $KAlSi_3O_8$, and albite, $NaAlSi_3O_8$, whereas at lower temperatures exsolution takes place, which produces more-or-less parallel alternating lamellae of the pure end-members. Feldspars having these exsolution lamellae are called *perthite*, *microperthite*, or *cryptoperthite*, according to

Orthoclase, $KAlSi_3O_8$

Crystallography. Monoclinic 2/*m*; crystals common, normally with prismatic habit and dominated by {010}, {110}, and {001}; can be elongated parallel to [001] and tabular parallel to {010}, or elongated along [100] with a squarer cross-section; also with pseudo-orthorhombic habit as in adularia. Twinning common: *Carlsbad twins* as penetration twins with the *c* axis as twin axis, *Manebach twins* with {001} as both twin and composition plane, and *Baveno twins* with {021} as twin and composition plane. Orthoclase occurs most commonly as grains in rocks or in more-or-less coarse-grained masses.

Physical properties. Cleavage perfect on {001}, good on {010}, and indistinct on {110}. Hardness 6; density 2.6. Colour is commonly weakly flesh-red, otherwise colourless, white, grey, yellowish, or greenish; sometimes with

Figure 566. Orthoclase from Vålerveien Quarry, Moss, Norway. Subject: 47 × 116 mm.

367

Figure 567. Orthoclase, with labradorescence, from Stavern (formerly Frederiksværn), Vestfold, Norway. Field of view: 90 × 135 mm.

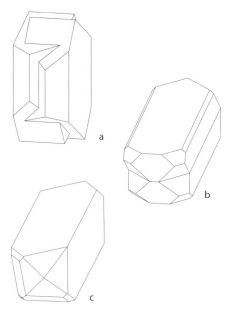

whether they are distinguishable by the naked eye, under the microscope, or are even finer. When albite lamellae dominate, the feldspar is called *antiperthite*. The crystal structure of orthoclase is similar to that of sanidine, except that in orthoclase Al and Si are partly ordered in the tetrahedral positions. This is related to the fact that orthoclase is formed at lower temperatures than sanidine.

Names and varieties. *Adularia* is a colourless and commonly transparent variety, typically found in single crystals. It occurs in hydrothermal veins, formed at low temperature. *Celsian*, $BaAl_2Si_2O_8$, is a related feldspar close-

Figure 568. Twinning in orthoclase: (a) Carlsbad twin, a penetration twin with the *c* axis as twin axis; (b) Manebach twin with {001} as twin and composition plane; (c) Baveno twin with {021} as twin and composition plane.

ly resembling orthoclase in structure and physical properties. Orthoclase and celsian form a solid-solution series in which *hyalophane* covers a range with K > Ba. Hyalophane is found in adularia-like crystals.

Occurrence. Orthoclase is the characteristic K-feldspar in igneous rocks such as granites and syenites, which are formed at lower temperatures than rocks with sanidine, and at higher temperatures than rocks and veins with microcline. Orthoclase also occurs in some metamorphic rocks and in certain sandstones.

Diagnostic features. Crystal habit, including twinning, cleavage, and mode of occurrence; orthoclase and sanidine do not conform to the twin laws for triclinic feldspars.

Microcline, $KAlSi_3O_8$

Crystallography. Triclinic $\bar{1}$; crystals common, habits normally as for orthoclase; twin laws for orthoclase also apply to microcline, but only Carlsbad twins are common; microcline is also found as twins on the laws for triclinic feldspars, the albite and pericline

Figure 569. Orthoclase, a Carlsbad twin, from Narssârssuk, Greenland. Subject: 42 × 70 mm.

Figure 570. Hyalophane from Bosna, Bosnia-Herzegovina. Subject: 55 × 62 mm.

Figure 571. Perthite from Perth, Ontario, Canada. Subject: 39 × 45 mm.

laws (described under albite); polysynthetic twinning on both twin laws commonly present together, resulting in the typical cross-hatched pattern seen under the microscope; exsolution lamellae (perthite) as described under orthoclase are also found in microcline. A special type of intergrowth of quartz and microcline is displayed in *graphic granite*. Microcline occurs mostly as grains in rocks and in coarse granular masses.

Figure 572. Graphic granite, an intergrowth of microcline and quartz, from Siberia, Russia. Subject: 48 × 87 mm.

Figure 573. Microcline var. amazonite from Florissant, Teller County, Colorado, USA. Subject: 64 × 56 mm.

Physical properties. Cleavage perfect on {001}, good on {010}. Hardness 6; density 2.6. Colour is white, pale yellow, reddish, or sometimes greenish (amazonite); vitreous lustre, pearly on cleavage surfaces; translucent.

Chemical properties, etc. Microcline normally contains some Na in the form of albite exsolution lamellae. The crystal structure of microcline is triclinic, differing from that of orthoclase primarily in a complete ordering of Al and Si in the tetrahedral positions. This is connected with the fact that microcline is formed at lower temperatures than orthoclase.

Names and varieties. *Amazonite* is a green variety that is popular for ornamental purposes.

Occurrence. Microcline is present in plutonic igneous rocks such as granites and syenites that have formed slowly and at great depth; it also occurs in gneisses and in some sandstones. It is the common K-feldspar in hydrothermal veins and pegmatites, in which it can be found as very large crystals, some weighing more than 2000 tonnes.

Use. Microcline from pegmatites is mined in large quantities and used in the production of porcelain, enamel, and glass.

Diagnostic features. Crystal habit, cleavage, and (partly) mode of occurrence; commonly seen as graphic granite; green feldspar is usually microcline.

Plagioclase series:
Albite, $NaAlSi_3O_8$ – anorthite, $CaAl_2Si_2O_8$

Crystallography. Triclinic $\bar{1}$; crystals mostly tabular parallel to {010}, rarely elongated along the *b* or *c* axis; twinning very common: twins on the Carlsbad, Manebach, and Baveno laws occur, but most common are twins on the *albite* and *pericline* laws. In albite twins {010} is the twin plane, and the twinning is usually polysynthetic, i.e. repeated in parallel lamellae like leaves in a book; lamellae are commonly visible with a hand lens and appear as grooves or striations, best studied on a {001} cleavage face. In pericline twins the *b* axis is the twin axis; these twins are also usually polysynthetic. Plagioclases are most common as grains in rocks and in more-or-less coarse-grained masses.

Physical properties. Cleavage perfect on {001}, good on {010}. Hardness 6; density 2.6

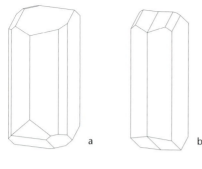

Figure 574. Albite:
(a) {010}, {001}, {110}, {1$\bar{1}$0}, {130}, {1$\bar{3}$0}, {10$\bar{1}$}, {20$\bar{1}$}, {0$\bar{2}$1}, {11$\bar{1}$}, and {$\bar{1}\bar{1}$2};
(b) albite twin with {010} as twin plane;
(c) repeated or polysynthetic twinning on the albite law.

Chemical properties, etc. At high temperature the plagioclases form an almost complete solid-solution series between albite and anorthite. At lower temperatures the miscibility is incomplete and various exsolution structures occur. They are not usually directly visible but can give rise to plays of colour. The miscibility in plagioclases takes place as a coupled substitution, because the substitution of Na^+ for Ca^{2+} has to be linked with a substitution of Al^{3+} for Si^{4+} in order to maintain charge balance. The chemical variation in plagioclases can thus be expressed by the equation $Na^+ + Si^{4+} \Leftrightarrow Ca^{2+} + Al^{3+}$.

A number of names are attached to a series of artificially chosen intervals in the plagioclase series, which can be expressed as percentages of the anorthite content: *albite* (An_0–An_{10}), *oligoclase* (An_{10}–An_{30}), *andesine* (An_{30}–An_{50}), *labradorite* (An_{50}–An_{70}), *bytownite* (An_{70}–An_{90}), and *anorthite* (An_{90}–An_{100}). The miscibility between plagioclases and alkali feldspars is limited to the albite end of the plagioclase series; as noted above, K- and Na-rich feldspars are frequently bracketed together as the *alkali feldspars*.

The crystal structure of plagioclase consists of a three-dimensional framework of [SiO_4] tetrahedra with spaces containing Na or Ca. The Si and Al ions are completely ordered in the tetrahedral positions.

Names and varieties. Besides the names defined under chemical properties a few more exist for plagioclases. *Peristerite* is a variety with a play of colour caused by exsolution lamellae, normally in the An_2–An_{20} interval. A play of colour of similar character is found in labradorite. *Moonstone* can be a plagioclase

(albite) to 2.8 (anorthite). Colourless, white, or grey, rarely greenish or reddish; a particular play of colour (labradorescence) is especially displayed by labradorite and andesine; vitreous lustre, pearly lustre on cleavage surfaces; mostly translucent.

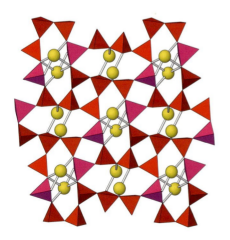

Figure 575. In albite the crystal structure consists of a three-dimensional network of [SiO_4] tetrahedra (red) and [AlO_4] tetrahedra (purple) with cavities containing the relatively large Na ions (yellow). Si and Al are completely ordered in the tetrahedral sites. The figure shows a slice of the structure viewed along the *a* axis, with the *b* axis E–W. Because only a slice is shown here, the nine bonds between Na and O (the tetrahedral corners) cannot be seen, nor is it evident that all corners of the tetrahedra are shared with other tetrahedra. (Compare with Figure 563.)

Figure 576. Albite from Narssârssuk, Greenland. Field of view: 17 × 27 mm.

with a bluish milky tinge in certain direc-tions (iridescence). *Clevelandite* is albite with a pronounced tabular crystal habit, and *aven-turine* is not only a quartz variety but also a plagioclase variety, often oligoclase, with in-clusions of hematite or mica, sometimes also called *sunstone*.

Occurrence. The plagioclases are the most important rock-forming minerals. In particu-lar, they constitute the principal part of all ig-neous rocks, and their systematic mode of oc-currence in these rocks has lead to a classifi-cation system of igneous rocks based on the chemical composition of plagioclases. Gener-ally speaking, the more Si-rich the rocks the higher the Na content in plagioclases; and the more Si-poor the rocks the higher the Ca content in plagioclases.

Albite is also found as well-formed crystals in pegmatites, whereas pure anorthite is a characteristic mineral of contact-metamor-phosed limestones.

Figure 577. Labradorite from an unknown locality. Field of view: 26 × 39 mm.

Figure 578. Oligoclase var. aventurine, striated by twin lamellae on the albite law, from Tvedestrand, Norway. Subject: 66×81 mm.

Figure 579. Albite var. peristerite from Hybla, Ontario, Canada. Subject: 85×88 mm.

Use. Plagioclase in pegmatites is used for similar purposes as K-feldspars. Labradorite and other plagioclases showing play of colour are used for ornamental work and as gemstones.
Diagnostic features. Crystal habit, cleavage, and mode of occurrence; twinning striation on a {001} cleavage face on the albite law distinguishes plagioclases from K-feldspars. The plagioclases cannot be distinguished from each other with certainty without chemical or optical tests.

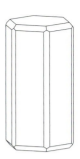

Figure 580. Nepheline: $\{10\bar{1}0\}$, $\{0001\}$, $\{11\bar{2}0\}$, and $\{10\bar{1}1\}$.

Feldspathoids

The feldspathoids are a group of tectosilicates containing Al that have common features with feldspars with respect to mode of occurrence and chemical composition, for which reason they are called *feldspathoids*, i.e. feldspar-like. The most important feldspathoids are *nepheline*, *leucite*, *sodalite*, and *cancrinite*. They differ from the feldspars primarily in having a lower Si content, and they typically occur in rocks that are rich in K or Na and poor in Si. Their crystal structures are generally more open than feldspars, and consequently they generally have a lower density.

Analcime, sometimes considered as a feldspathoid, is described with the zeolites.

Nepheline, (Na,K)AlSiO$_4$

Crystallography. Hexagonal 6; crystals rare, typically small simple crystals consisting of prism and pinacoid; mostly as grains in rocks or in irregular aggregates.
Physical properties. Cleavage indistinct on $\{10\bar{1}0\}$. Hardness 6; density 2.6. Colourless, white, or grey, sometimes with a brownish, greenish, or reddish tinge; greasy lustre; transparent to translucent.

Figure 581. Nepheline from Larvik, Norway. Subject: 75 × 95 mm.

Chemical properties, etc. The Na/K ratio varies but is usually close to 3:1.

Names and varieties. *Kalsilite*, $KAlSiO_4$, is a closely related mineral occurring in alkali-rich basalts.

Occurrence. Nepheline is a rock-forming mineral in nepheline-syenites and similar igneous rocks poor in Si, where it replaces feldspars. Significant occurrences are known, e.g. from the Ilímaussaq and Igaliko complexes in Greenland and at Langesundsfjorden, Norway. The largest occurrence of nepheline is in the Khibina massif in Kola Peninsula, Russia, where it is mined together with apatite. Large crystals are known from Davis Hill in Ontario and fine crystals from Mont Saint-Hilaire, Quebec, Canada.

Use. Nepheline is used locally in the glass industry; in the Kola Peninsula it has been used for many purposes, e.g. in the production of ceramics, leather, textiles, and rubber.

Diagnostic features. Greasy lustre and, in comparison with quartz, a lower hardness.

Leucite, $K(AlSi_2)O_6$

Crystallography. Cubic $4/m\overline{3}2/m$ above 605 °C, tetragonal $4/m$ below; crystals mostly in the cubic form {211}, which in the low-temperature phase consists of tetragonal twin lamellae.

Physical properties. No cleavage. Hardness 6; density 2.5. Colour is most often dull white or grey, rarely colourless; vitreous lustre; translucent.

Chemical properties, etc. Small amounts of K can be replaced by Na.

Occurrence. Leucite is a characteristic mineral in young Si-poor extrusive igneous rocks related to the Mediterranean igneous province; it is found, e.g. as crystals in the lavas from Vesuvius, Italy.

Diagnostic features. Crystal form and mode of occurrence; leucite resembles analcime in crystal habit, but is normally embedded in a fine-grained matrix, whereas analcime is found crystallized in cavities.

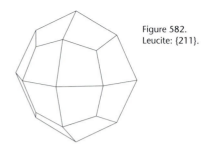

Figure 582. Leucite: {211}.

Sodalite, $Na_4(Si_3Al_3)O_{12}Cl$

Crystallography. Cubic $\overline{4}3m$; crystals rare, usually with {110}; mostly as grains in rocks or massive.

Physical properties. Cleavage indistinct on {110}. Hardness 6; density 2.3. Colour is usually grey, bluish, greenish, or reddish, also white; vitreous lustre; translucent or transparent.

Chemical properties, etc. Sodalite has an open crystal structure with very large spaces, in which large complex anions such as Cl^-, S^{2-}, or SO_4^{2-} can be accommodated.

Names and varieties. *Haüyne* and *nosean* are closely related minerals with SO_4 instead of Cl; they are primarily found in young extrusive igneous rocks.

Occurrence. Sodalite is a rock-forming mineral in some alkaline rocks, such as sodalite-syenites, nepheline-syenites, trachytes, and phonolites, commonly in association with nepheline or cancrinite. It is found, e.g. in the Ilímaussaq complex in Greenland, where it has a reddish colour when fresh but after a few seconds changes to a light greenish colour. Exceptional crystals are found at Mont Saint-Hilaire, Quebec, Canada.

Diagnostic features. Colour and changes in colour can be characteristic, but sodalite can otherwise be difficult to distinguish from other feldspathoids.

Figure 583. Leucite from Vesuvius, Italy. Field of view: 58 × 83 mm.

Figure 584. In the crystal structure of sodalite minerals a three-dimensional network of $[SiO_4]$ tetrahedra (red) and $[AlO_4]$ tetrahedra (purple) contains large cavities, in which there is room for large anions or complex anions such as Cl^-, S^{2-}, or SO_4^{2-} (not shown).

Figure 585. Sodalite with eudialyte and arfvedsonite from the Ilímaussaq complex, Greenland. Subject: 64 × 84 mm.

Figure 586. Blue sodalite from Hondoto River, Swartbooisdrift, Namibia. Subject: 68 × 85 mm.

Figure 587. Lazurite from Badakhshan, Afghanistan.
Subject: 41 × 68 mm.

Lazurite, $Na_3Ca(Si_3Al_3)O_{12}S$

Crystallography. Cubic $\bar{4}3m$; crystals rare, mostly {110}; occurs primarily massive.

Physical properties. Cleavage indistinct on {110}. Hardness 5-6; density 2.4. The colour is typically azure-blue, rarely greenish-blue, often variable in intensity; streak light blue; vitreous lustre; translucent.

Chemical properties etc. Lazurite has a sodalite crystal structure. S^{2-} is in variable amounts replaced by Cl^- or SO_4^{2-}.

Names and varieties. *Lapis lazuli* is more-or-less synonymous with lazurite; the name is also used for a mixture of lazurite with minor amounts of other minerals, including pyrite.

Occurrence. Lazurite is rather rare; it is especially found in contact-metamorphic limestones. The best-known deposits are in the Badakhshan Province in Afghanistan and at Lake Baikal in Russia.

Use. Lazurite has from the earliest times been prized as material for gemstones, *objets d'art*, wall decoration, etc. The pigment ultramarine was in the past made from powdered lazurite.

Diagnostic features. Colour; pyrite is frequently an associated mineral.

Tugtupite, $Na_4BeAlSi_4O_{12}Cl$

Tetragonal; it is related to sodalite and has the same crystal structure except for having (BeSi) instead of (AlAl) as in sodalite. Tugtupite has properties almost identical with those of sodalite, except for a lower hardness; it is mostly found in granular masses that can be white, pink, or carmine. Colour often varies in intensity; sometimes becomes more intense on exposure of sunlight and bleached when stored in the dark. Tugtupite is primarily known from the Ilímaussaq complex in Greenland, where it is found in hydrothermal veins in association with, e.g. albite, analcime, and aegirine. As a gemstone, it is mostly cut *en cabochon*.

Figure 588. Tugtupite from Kvanefjeld, Ilímaussaq complex, Greenland. Subject: 101 × 110 mm.

Figure 589. Cancrinite from Litchfield, Maine, USA. Subject: 81 × 129 mm.

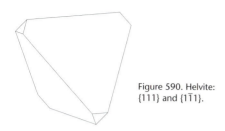

Figure 590. Helvite: {111} and {1$\bar{1}$1}.

Cancrinite,
$(Na,Ca)_8(Si_6Al_6)O_{24}(CO_3)_2 \cdot 2H_2O$

Hexagonal, found in prismatic crystals but occurs mostly massive. cleavage on {10$\bar{1}$0}; hardness 5–6; density about 2.5. The colour varies, most often yellowish; vitreous lustre; translucent. Cancrinite is found in alkaline rocks such as sodalites and nepheline-syenites. It is known, e.g. from Litchfield in Maine, USA.

Helvite, $Be_3Mn_4(SiO_4)_3S$

Cubic $\bar{4}3m$; commonly in crystals with {111} and {1$\bar{1}$1}; also massive. Cleavage on {111} distinct; hardness 6; density about 3.3. Colour is usually yellowish; vitreous lustre; translucent. Helvite forms solid-solution series with *danalite*, $Be_3Fe_4(SiO_4)_3S$, and *genthelvite*, $Be_3Zn_4(SiO_4)_3S$. These minerals are found in granitic and in nepheline-syenite pegmatites, in skarn deposits, and in hydrothermal veins. Helvite is known, e.g. from Butte in Montana and Amelia in Virginia, USA.

Scapolite

Solid-solution series exist between *marialite*, $(Na,Ca)_4(Si,Al)_{12}O_{24}(Cl,CO_3,SO_4)$, and *meionite*, $(Ca,Na)_4(Si,Al)_{12}O_{24}(CO_3,SO_4,Cl)$

Crystallography. Tetragonal 4/m; crystals common, commonly in combinations of {100}, {110}, and {111} with no or only minor {001}; crystals commonly rough with uneven faces; also in granular masses.

Physical properties. Cleavage distinct on {100} and {110}. Hardness 5–6; density 2.5–2.7. Colour is white, grey, or light greenish, rarely yellowish, bluish, or reddish; vitreous lustre; translucent to transparent.

Chemical properties, etc. There is a complete solid-solution series between marialite and meionite. In its chemistry this series corresponds to the plagioclase series, with the difference that the more open crystal structure of scapolite can accommodate large anions, whether simple or complex.

Figure 591. Helvite on feldspar from Kangerdluarsuk, Ilímaussaq complex, Greenland. Subject: 24 × 30 mm.

Figure 592. Scapolite:
{100}, {001}, {110},
and {111}.

It is also a characteristic mineral in contact-metamorphic limestones, where it can occur locally in larger masses. Scapolite of gemstone quality is found, e.g. in Sri Lanka and the Mogok district in Myanmar.

Diagnostic features. Crystal habit and cleavage; massive forms resemble feldspar but usually have a more splintered appearance owing to the cleavage.

Occurrence. Scapolite is a common mineral in regional metamorphic rocks such as gneisses and amphibolites, where it is commonly formed at the expense of plagioclase.

Ussingite, $Na_2AlSi_3O_8(OH)$

Triclinic; mostly in granular masses. Cleavage distinct on {001}, indistinct on {110}; hardness 6½; density 2.5. Colour is reddish-violet to white; vitreous lustre; mostly transparent. Ussingite is found in nepheline-syenite pegmatites, e.g. in the Ilímaussaq complex in Greenland and in the similar occurrences in the Kola Peninsula, Russia. Fine crystals are found at Mont Saint-Hilaire, Quebec, Canada.

Zeolites

The zeolites are a large group of water-containing tectosilicates characterized by particularly open crystal structures. They have a three-dimensional framework of $[SiO_4]$ and $[AlO_4]$ tetrahedra that encloses open cavities in the form of channels and cages that give space for H_2O molecules and, e.g. Na^+, Ca^{2+}, and K^+.

The H_2O is loosely bound in the structure, and on heating it is continuously given off without causing the structure to collapse. H_2O can subsequently be absorbed and replaced by suspending the specimen in water. Large cations such as Na^+ can also be given off; this results in

Figure 593. Scapolite from Arendal, Norway. Field of view: 47 × 81 mm.

Figure 594. Ussingite from Kangerdluarsuk, Ilímaussaq complex, Greenland. Subject: 74 x 94 mm.

Figure 595. Analcime: {211}.

a deficit in the charge balance, which can be compensated by taking up other cations, e.g. K^+ or $\frac{1}{2}Ca^{2+}$. In other words, zeolites are capable of exchanging ions, a property that is of practical value, e.g. for softening and cleaning water. The ability to exchange ions, although facilitated by the loose bonding of the ions in the structure, depends more on the

Figure 596. Analcime from the Alpe di Siusi near Bolzano, Italy. Subject: 88 x 85 mm.

383

simple physical fact that the routes for transporting the ions, i.e. the channels, are sufficiently large to let them pass. In situations where cations of some types can be absorbed and others cannot because they are too large, a zeolite can function as a molecular sieve, a property that is extensively exploited in industry. Almost all zeolites used today as molecular sieves are produced synthetically, and it is now possible to create a zeolite for a particular purpose by designing it with the required mesh dimensions. Natural zeolites are used in the more basic applications like soil conditioning, where large volumes are required, whereas synthetic zeolites are restricted to more specific applications.

Compared with other tectosilicates, zeolites have a lower hardness and a lower density. This is a consequence of the open structure. They are colourless, white, or only pale coloured, and occur as needle-shaped or fibrous crystals, as tabular crystals, or as cube-like crystals.

Zeolites occur as well-crystallized fillings in cavities in basalts and other extrusive igneous rocks. They are also important constituents in sedimentary deposits, e.g. in the western USA, where zeolites are formed by alteration of layers of volcanic ash.

Analcime, $Na(AlSi_2)O_6 \cdot H_2O$

Crystallography. Cubic $4/m\bar{3}2/m$; crystals usually as {211}; also granular.
Physical properties. No cleavage. Hardness 5–5½; density 2.3. Colourless or white, rarely greyish, yellowish, or reddish; vitreous lustre; translucent or translucent.

Chemical properties, etc. Small amounts of Na can be replaced by K. The crystal structure of analcime has large channels with H_2O, as in the zeolites. It is now classified as a member of the zeolite group.
Occurrence. Analcime is found as a primary rock-forming mineral in igneous rocks as well as a hydrothermal product in veins and cavities in rocks; in the latter case usually as well-developed crystals. Large crystals are known from Mont Saint-Hilaire in Quebec, Canada.
Diagnostic features. Crystal habit and mode of occurrence; analcime crystals differ from leucite by not being embedded in a fine-grained matrix.

Pollucite, $(Cs,Na)(AlSi_2)O_6 \cdot nH_2O$, is a rare cubic mineral mostly found in aggregates resembling feldspar or milky quartz in pegmatites, e.g. in Varuträsk in Sweden. It is now classified as a zeolite.

Natrolite, $Na_2(Si_3Al_2)O_{10} \cdot 2H_2O$

Crystallography. Orthorhombic $mm2$; crystals common, often long prismatic to needle-shaped and in radiating aggregates; also fibrous, granular.
Physical properties. Cleavage perfect on {110}. Hardness 5–5½; density 2.3. Colourless or white, more rarely yellowish or reddish; vitreous lustre; transparent to translucent.
Chemical properties, etc. Na can in small amounts be replaced by Ca or K. In the three-dimensional framework of $[SiO_4]$ and $[AlO_4]$ tetrahedral chains run parallel to the c axis; between the chains are cavities in which Na and H_2O are present. Na is bonded to six O atoms, four from chains and two from the H_2O molecules.
Occurrence. Natrolite occurs in cavities in basalts, commonly in association with other zeolites. Particularly beautiful groups of natrolite crystals are found at several localities in northern Bohemia, Czech Republic. Natrolite is also found in nepheline-syenites and other igneous rocks, where it occurs either as a late

Figure 597. Natrolite: {100}, {010}, {110}, {310}, and {111}.

Figure 599. Natrolite from the former Neubauerberg near Ceska Lipá, Czech Republic. Field of view: 36 × 54 mm.

primary mineral or as a secondary alteration product of nepheline.

Diagnostic features. Mode of occurrence and crystal habit; fibrous zeolites are otherwise generally very similar in appearance.

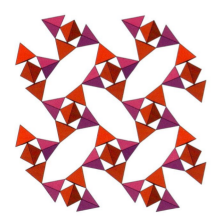

Figure 598. The crystal structure of natrolite, seen along the *c* axis, consists of a three-dimensional network of [SiO$_4$] tetrahedra (red) and [AlO$_4$] tetrahedra (purple) forming chains along the *c* axis; between the chains, cavities with Na and H$_2$O (not shown) are present; Na is bonded to six oxygen atoms, four from the chains and two from the H$_2$O molecules.

Scolecite, $Ca(Si_3Al_2)O_{10} \cdot 3H_2O$, and mesolite, $Na_2Ca_2(Al_6Si_9)O_{30} \cdot 8H_2O$

Both minerals are monoclinic and are closely related to natrolite. In general, their physical properties are like those of natrolite, with a fibrous or needle-shaped crystal habit; mesolite, however, is often more hair-like, and scolecite usually has thicker needle- or ruler-shaped crystals. They occur, like natrolite, in cavities in basaltic lavas, but are also known from cavities in granites and syenites. Teigarhorn in Berufjord, Iceland, is a well-known locality for scolecite; mesolite is especially known from the Faeroe Islands. Both minerals are found as exceptionally fine crystals in Poona, India.

Figure 600. Scolecite from Teigarhorn, Iceland. Subject: 57 × 58 mm.

Gonnardite, $(Na,Ca)_2(Si,Al)_5O_{10} \cdot 3H_2O$

Orthorhombic; also closely related to natrolite, both structurally and in having similar physical properties. It occurs primarily in Si-poor igneous rocks and related pegmatites and is known, e.g. from Crestmore in California, USA.

Thomsonite, $NaCa_2(Al_5Si_5)O_{20} \cdot 6H_2O$

Orthorhombic; crystals uncommon; variable habits; mostly in fan-shaped or spherical aggregates. Cleavage perfect on {010}; hardness 5–5½; density 2.3. Colour is white, grey, yellow, or red; vitreous lustre, pearly lustre on cleavage surfaces; mostly translucent. Thomsonite is a common zeolite in cavities in phonolites and basalts; it is also found in cavities in nepheline-syenites. It is known, e.g. from the basalts in Scotland, the Faeroe Islands, and Iceland.

Mordenite, $K_{2.8}Na_{1.5}Ca_2(Al_9Si_{39})O_{96} \cdot 29H_2O$

Orthorhombic; occurs partly as white needle-shaped crystals, commonly in radiated aggregates, partly in compact porcellaineous masses with variable colour. It occurs like other zeolites in cavities in lavas, but is also known from sediments formed by the alteration of layers of volcanic ash.

Figure 601. Mesolite from Midvaag harbour, Vågø, the Faeroe Islands. Field of view: 70 × 90 mm.

Figure 602. Gonnardite from Island Magee, Antrim, Northern Ireland. Subject: 61×70 mm.

Figure 603. Thomsonite from Catania, Palagonia, Sicily, Italy. Field of view: 48×72 mm.

Figure 604. Mordenite from Teigarhorn, Iceland. Subject: 81 × 109 mm.

Laumontite, $Ca(Al_2Si_4)O_{12}\cdot4H_2O$

Crystallography. Monoclinic $2/m$; crystals commonly long prismatic with predominant {110}; twinning on {100} seen; also massive.
Physical properties. Cleavage perfect on {010} and {110}: hardness 3–4; density about 2.3. White or colourless, rarely pink; vitreous lustre, pearly lustre on cleavage surfaces, often chalk-like when dehydrated; translucent.
Chemical properties, etc. In normal storage laumontite readily loses part of its water con-

Figure 605. Laumontite from North Island, New Zealand. Subject: 59 × 79 mm.

Figure 606. Heulandite: {100}, {010}, {001}, {$\bar{1}$01}, {021}, {22$\bar{1}$}, and {22$\bar{3}$}.

tent and is transformed into a chalk-like brittle material sometimes called *leonhardite*.

Occurrence. Laumontite occurs in a number of different associations, e.g. in 'zeolite-facies' rocks (i.e. metamorphic rocks altered under low-pressure–temperature conditions); in sediments, where it is formed by alteration of volcanic glass, plagioclase, etc.; and as fillings in veins and cavities in igneous rocks.

Diagnostic features. The simple crystal habit and tendency to dehydrate.

Heulandite
$(Na,K)Ca_4(Al_9Si_{27})O_{72}\cdot24H_2O$

Crystallography. Monoclinic $2/m$; crystals common, usually somewhat tabular parallel to {010} and with an orthorhombic-like development of other faces.

Physical properties. Cleavage perfect on {010}. Hardness 3½–4; density 2.2. Colourless or white, also yellowish or reddish owing to impurities; vitreous lustre, pearly lustre on cleavage surfaces; transparent to translucent.

Chemical properties, etc. The chemical formula is also written in a more complex form with Ca, Sr, Ba, and Mg. (Heulandite-Ca, heulandite-K, heulandite-Na, and heulandite-Sr are recognized as separate species.) The Na/Ca and Si/Al ratios can vary considerably.

Occurrence. Heulandite occurs in cavities in basaltic rocks, commonly associated with stilbite and other zeolites. Particularly beautiful specimens are found in Iceland and the Faeroe Islands.

Diagnostic features. Crystal habit, cleavage, and lustre.

Figure 607. Heulandite from Sayad Pimpri Mine, Nasik, Bombay, India. Field of view: 60×87 mm.

Clinoptilolite
$(Na,K,Ca)_{2-3}(Si_{15}Al_3)O_{36}\cdot11H_2O$

Monoclinic; structurally closely related to heulandite and has largely the same physical properties. Chemically it is characterized by a higher Si content than in heulandite, and also by a considerable variation in the ions in the cavities, since Na and K, as well as Ca, can be predominant. Clinoptilolite is a widespread mineral, which together with phillipsite and other zeolites constitutes the bulk of the red deep-sea sediments. It is also common in other sediments as an alteration product of Si-rich volcanic ash; such deposits are especially known from the western USA, where in several places clinoptilolite occurs in large amounts. It has good ion-exchange properties and is used for many purposes, e.g. removing radioactive Cs and Sr isotopes from waste water from nuclear reactors.

Figure 608. Heulandite from Iceland. Subject: 33 × 75 mm.

Stilbite, $NaCa_4(Al_9Si_{27})O_{72} \cdot 30H_2O$

Crystallography. Monoclinic 2/*m*; crystals common, normally assembled in characteristic sheaf-like aggregates.

Physical properties. Cleavage perfect on {010}. Hardness 3½–4; density 2.2. Colourless, white, grey, yellowish, or brownish; vitreous lustre, pearly lustre on {010}; transparent to translucent.

Chemical properties etc. Ca can be partly replaced by Na or K.

Names and varieties. *Desmine* is an obsolete name for stilbite.

Occurrence. Stilbite is a common mineral in cavities in basalts, where it is commonly found in association with heulandite and other zeolites. Well-known occurrences include the basalts of Iceland, the Faeroe Islands, and Scotland, and Poona, India.

Diagnostic features. The sheaf-like aggregate form.

Figure 609. Stilbite from Iceland. Field of view: 70 × 96 mm.

Phillipsite, $K(Ca_{0.5},Na)_2(Si_5Al_3)O_{16}\cdot 6H_2O$

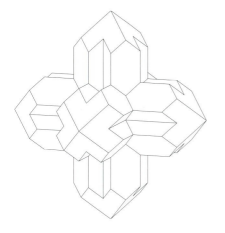

Monoclinic; found as more-or-less complicated penetration twins simulating orthorhombic, tetragonal, or cubic symmetry. Cleavage distinct on {010}, indistinct on {100}; hardness 4½; density 2.2. Colourless, white or pale yellowish; vitreous lustre; transparent to translucent. The content of K, Ca, and Na varies, and Ba and Sr can to a limited extent substitute for these ions. Phillipsite is a common mineral in deep-sea sediments; it also occurs in cavities in basalts in association with other zeolites.

Figure 610. Phillipsite: A group of twins simulating cubic symmetry. Three fourlings stand perpendicular to each other as fourfold axes in a rhombic dodecahedron; each fourling consists of two sets of penetration twins.

Figure 611. Phillipsite from Palagonia, Italy. Field of view: 24 × 36 mm.

Figure 612. Harmotome from Strontian, Scotland. Field of view: 36 × 54 mm.

Harmotome, $BaAl_2Si_6O_{16} \cdot 6H_2O$

Monoclinic; found as penetration twins as for phillipsite. The physical properties are almost the same as for phillipsite, but the density of about 2.5 is higher owing to the Ba content. Harmotome is typically found in late hydrothermal veins, and is known, e.g. from St. Andreasberg in the Harz Mountains, Germany, and Strontian in Scotland

Chabazite, $Ca(Al_2Si_4)O_{12} \cdot 6H_2O$

Crystallography. Trigonal $\bar{3}2/m$; crystals common, often as cube-like rhombohedra $\{10\bar{1}1\}$; penetration twins on $\{0001\}$ frequent. **Physical properties.** Cleavage indistinct on $\{10\bar{1}1\}$. Hardness 4–5; density 2.1. Colourless or white, also yellowish to pink; vitreous lustre; transparent to translucent.

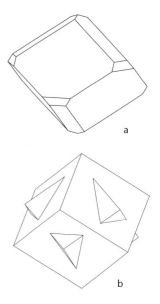

Figure 613. Chabazite: (a) $\{10\bar{1}1\}$, $\{01\bar{1}2\}$, and $\{02\bar{2}1\}$; (b) penetration twin on $\{0001\}$.

Figure 614. Chabazite from Qeqertarsuaq, Greenland. Field of view: 42 × 58 mm.

Figure 615. Gmelinite from Larne, Ireland. Field of view: 40 × 60 mm.

Chemical properties, etc. K, Sr, and Na can to some degree replace Ca.

Occurrence. Chabazite typically occurs with other zeolites in cavities in basalts and andesites. It is also widespread in volcanic tuffs, where it is formed by disintegration of Ca-rich plagioclase. Fine chabazite crystals are known, e.g. from Paterson in New Jersey, USA.

Diagnostic features. Crystal habit; differs from calcite in having a greater hardness and in not effervescing in hydrochloric acid.

Gmelinite, $Na_4(Al_4Si_8)O_{24}\cdot11H_2O$

Hexagonal, found as simple, somewhat tabular, crystals with $\{10\overline{1}0\}$, $\{10\overline{1}1\}$, and $\{0001\}$. Cleavage good on $\{10\overline{1}0\}$; hardness 4½, density 2.1. Colourless or white, also light yellowish or reddish; vitreous lustre; mostly translucent. Gmelinite is found both in cavities in basalts and in Na-rich veins in alkaline rocks. It commonly occurs in oriented intergrowths with chabazite.

Organic minerals

Among organic compounds in the geological environment a few well-defined crystalline minerals are found, as well as a group of poorly defined hydrocarbons such as asphalt, coal, etc. that have such a great variation in chemical composition that they must be considered as rocks. According to this distinction, amber is not, strictly speaking, a mineral, but for reasons of tradition it is included.

Whewellite, $CaC_2O_4 \cdot H_2O$

Monoclinic; found as small well-developed prismatic crystals, commonly in heart-shaped twins on $\{\bar{1}01\}$. Cleavage good on $\{\bar{1}01\}$, conchoidal fracture; hardness 2½, density 2.2. Colourless, yellowish, or brownish; pearly lustre. Whewellite occurs especially in association with coal seams, as near Dresden in Germany, but is also found as a primary mineral in veins, e.g. with tetrahedrite and quartz at Urbes in Alsace, France.

Figure 616. Amber from the North Sea, about 250 km west of Harboøre, Denmark. The amber lump weighs 1749 g. Subject: 101 × 158 mm.

Mellite, $Al_2C_6(COO)_6 \cdot 16H_2O$

Tetragonal; crystals rare, mostly nodular or as coatings. No distinct cleavage, conchoidal fracture; hardness 2–2½, density 1.6. Honey-coloured, also reddish or brownish; greasy vitreous lustre, transparent. Mellite occurs in fissures in brown coal and lignite.

Amber

Consists of C, H, and O in variable proportions, amorphous; occurs in lumps. Conchoidal fracture, brittle; hardness 2–2½, density 1.1. Colour is yellowish, brownish, or whitish; resinous lustre; transparent to translucent; becomes electrified by rubbing; melts readily and burns with a pale, sooty flame.

Amber is a resin from extinct coniferous trees that grew in the Baltic region during Tertiary times. Some of the original layers, subsequently hardened by time, were later re-deposited by glaciers and rivers and in part transported westwards to Poland, northern Germany, and Denmark.

In prehistoric times amber was an important commodity in the countries around the Baltic Sea. It was at that time, and still is, a popular material for trinkets, etc.

TABLES

Tables: common minerals and their properties

Systematic determinative tables based on genetic relations, such as are used in botany, are not available in mineralogy. We have to observe as many properties of the specimen as possible and then try to identify it by comparing our observations with the available mineral descriptions. As an aid to this identification puzzle, some of the most common minerals are listed below in two tables arranged according to properties that are relatively easy to determine.

Table 1 includes minerals having a metallic or submetallic lustre; Table 2 includes minerals with a vitreous lustre or lustre of another type that is non-metallic. A few minerals have a lustre that makes it difficult to decide which table they belong to. They are therefore included in both tables. The minerals are ordered according to increasing hardness and in intervals of equal hardness according to increasing density. Since this arrangement is highly schematic it is recommended to search not only among minerals fitting the observations perfectly or nearly perfectly but also in the adjacent entries. Once a number of possible candidates have been isolated, one can go back to consult the more detailed descriptions in the more systematic Part 2.

Explanatory notes for Tables 1 and 2

Mineral, the name and formula with reference to the description in Part II of the text.

cs, the crystal system. A, triclinic; M, monoclinic; O. orthorhombic; Q, tetragonal; T, trigonal; H, hexagonal; C, cubic.

Habit, the typical habit including common crystal forms and types of aggregates; xl. is an abbreviation for crystal.

Cleavage, the most prominent directions of cleavage; occasionally also information on parting and fracture.

Brittleness, etc., properties such as brittleness, malleability, sectility, elasticity, magnetism, radioactivity, etc.

H, the hardness on Mohs' scale.

D, density.

Colour, the most common colours.

Streak (Table 1 only), the streak.

Lustre (Table 2 only), the lustre.

Table 1. Minerals with metallic or submetallic lustre, arranged according to hardness and density

Mineral	Page	cs	Habit	Cleavage	Brittle-ness etc.	H	D	Colour	Streak
Graphite, C	92	H	xl. rare, foliated, granular	perfect {0001}	flexible, inelastic, greasy	1	2.2	black	black
Pyrolusite, MnO_2	170	Q	xl. uncommon, radiated, fibrous, earthy	perfect {110}		1-2, also >2	4.4-5.1	steel-grey, iron-black, bluish tinge	black
Covellite, CuS	110	H	xl. uncommon, massive	perfect {0001}	slightly flexible	1½-2	4.7	blue to black etc.	black, shiny
Sylvanite, $AgAuTe_4$	123	M	xl. as characters	perfect {010}	brittle	1½-2	8.2	silver-white	grey
Stibnite, Sb_2S_3	113	O	xl. common, prismatic, needle-shaped, radiated	perfect {010}	inelastic, striated	2	4.6	lead-coloured, black, tarnishes	lead-coloured, black
Acanthite, Ag_2S	94	M	xl. rare, arborescent, massive	no distinct	flexible, sectile	2-2½	7.3	black, tarnishes dull	black, shiny
Bismuth, Bi	87	T	xl. rare, foliated, lamellar, massive	perfect {0001}	slightly brittle	2-2½	9.7	silver-white, reddish tinge	silver-white
Galena, PbS	97	C	xl. common, {100}, {111}, {110}, massive	perfect {100}		2½	7.6	lead-coloured, shiny to dull	lead-coloured
Chalcocite, Cu_2S	95	M	xl. rare, tabular, massive	indistinct {110}	slightly brittle	2½-3	5.7	lead-coloured, tarnishes	grey, black
Bournonite, $CuPbSbS_3$	127	O	xl. {001}, massive, granular	indistinct {010}		2½-3	5.8	steel-grey, black	black
Boulangerite, $Pb_5Sb_4S_{11}$	128	M	xl. rare, fibrous	good {100}	brittle, fibre flexible	2½-3	6	lead-coloured, bluish tinge	brownish grey
Copper, Cu	82	C	xl. rare, massive, dendritic	no	malleable	2½-3	8.9	copper-red, tarnishes	red, shiny
Silver, Ag	80	C	xl. rare, wire-like, massive	no	malleable	2½-3	10.5	silver-white, tarnishes	silver-white
Gold, Au	77	C	xl. rare, scaly, nuggets	no	malleable, sectile	2½-3	19.3	yellow	shiny yellow

Mineral	Page	cs	Habit	Cleavage	Brittleness etc.	H	D	Colour	Streak
Enargite, Cu_3AsS_4	127	O	xl. common, {001}, tabular, granular, columnar	perfect {110}, also {100}, {010}	brittle	3	4.5	steel-grey, black, tarnishes dull	black
Bornite, Cu_5FeS_4	96	Q	xl. rare, granular, massive	no distinct		3	5.1	bronze-yellow, tarnishes bluish-red	greyish black
Millerite, NiS	109	T	xl. prismatic or needle-shaped, radiated, fibrous	perfect {10$\bar{1}$1}, {01$\bar{1}$2}	brittle, needles flexible, elastic	3-3½	5.5	light brass-coloured	black, greenish tinge
Antimony, Sb	86	T	xl. rare, granular, lamellar	perfect {0001}	brittle	3-3½	6.7	tin-white to grey	lead-coloured, shiny
Tetrahedrite, $(Cu,Fe)_{12}$ Sb_4S_{13}	126	C	xl. {111}, massive, granular	no		3-4	4.8	grey, black	brown, black
Arsenic, As	86	T	xl. rare, scaly, botryoidal	perfect {0001}	brittle	3½	5.7	tin-white, tarnishes	grey
Chalcopyrite, $CuFeS_2$	104	Q	xl. e.g. {112}, massive	no distinct	brittle	3½-4	4.2	brass-coloured	greenish black
Pentlandite, $(Ni,Fe)_9S_8$	109	C	xl. very rare, granular	{111}	brittle	3½-4	4.8	light bronze-yellow	brownish
Cuprite, Cu_2O	145	C	xl. common, {111}, {100}, {110}, massive, earthy	distinct {111}	brittle	3½-4	6.1	red, black	reddish brown, lustre sub-adamantine
Manganite, MnO(OH)	179	M	xl. rare, prismatic, columnar, fibrous	perfect {010}, also {110}, {001}		4	4.3	steel-grey, black	dark brown, black, lustre submetallic
Stannite, Cu_2FeSnS_4	105	Q	granular	no distinct		4	4.4	steel-grey, black	black
Pyrrhotite, $Fe_{1-x}S$	106	H, M	xl. uncommon, {0001}, massive, granular	no distinct	brittle, magnetic	4	4.6	bronze-yellow, brown	greyish black
Platinum, Pt	83	C	xl. rare, grains or nuggets	no	malleable	4-4½	21.5	steel-grey, dark grey	whitish grey, bright
Iron, Fe	84	C	no xl., granular or lamellar	poor {100}	malleable, magnetic	4½	7-8	grey, black	grey, bright

Mineral	*Page*	cs	Habit	Cleavage	Brittle-ness etc.	H	D	Colour	Streak
Wolframite, $(Fe,Mn)WO_4$	*223*	M	xl. {hk0}, {100}, striated, granular, lamellar	perfect {010}		5	7.1-7.5	brownish black, black	reddish brown, black
Goethite, $FeO(OH)$	*180*	O	xl. rare, prismatic, {010}, massive, fibrous, earthy	perfect {010}, good {100}		5-5½	3.3-4.4	dark brown, yellowish brown, reddish brown	yellowish brown, lustre variable, dull
Aeschynite, (Ce,Ca,Fe) $(Ti,Nb)_2$ $(O,OH)_6$	*173*	O	xl. uncommon, prismatic, massive		metamict	5-6	4.2-5.3	dark brown	brown to black
Romanèchite, $(Ba,H_2O)_2$ Mn_5O_{10}	*171*	O	crusts, earthy			5-6	4.7	black, brownish black	black, brownish black
Uraninite, UO_2	*175*	C	xl. {100}, {111}, massive, banded	no	radioactive	5-6	6.5-11	black, brownish black	brownish black, lustre pitchy, dull
Fergusonite, $YNbO_4$	*174*	Q	xl. prismatic, granular		metamict	5-6½	4.2-5.7	black, brownish black	lustre submetallic
Perovskite, $CaTiO_3$	*164*	O	xl. cube-like, granular	no distinct		5½	4.0	black, brown, yellow	grey, white, lustre subadamantine
Chromite, $FeCr_2O_4$	*155*	C	xl. rare, granular, massive	no		5½	4.6	black	brown
Cobaltite, $CoAsS$	*120*	O	xl. {100}, {hk0}, granular	cube-like, variable	brittle	5½	6.3	silver-white, reddish tinge	greyish black
Nickeline, $NiAs$	*107*	H	xl. uncommon, tabular, massive	no distinct		5½	7.8	light copper-red, tarnishes	brownish black
Allanite, $Ca(Ce,La)$ $(Al,Fe)_3(Si_2O_7)$ $(SiO_4)(O,OH)_2$	*285*	M	xl. uncommon, granular, massive	indistinct	metamict, radioactive	5½-6	3.5-4.2	pitchy black, dark brown	dark brown, lustre greasy vitreous
Brookite, TiO_2	*172*	O	xl. {010} tabular, {120} prismatic	indistinct {120}		5½-6	4.1	yellowish brown, reddish brown, black	lustre subadamantine
Hausmannite, Mn_3O_4	*155*	Q	xl. uncommon, massive, granular	perfect {001}		5½-6	4.8	brown to black	brownish

Mineral	Page	cs	Habit	Cleavage	Brittle-ness etc.	H	D	Colour	Streak
Arseno-pyrite $FeAsS$	120	M	xl. common, prismatic, granular	distinct {001}		5½-6	6.1	silver-white to steel-grey	nearly black
Skutter-udite, $(Co,Ni)As_3$.	123	C	xl. uncommon, granular, massive	no distinct	brittle	5½-6	6.5	tin-white, steel-grey	black
Pyrolusite, MnO_2	170	Q	xl. uncommon, radiated, fibrous, earthy	perfect {110}		6-6½, also <6	4.4-5.1	steel-grey, iron-black, bluish tinge	black
Ilmenite, $FeTiO_3$	163	T	xl. uncommon, tabular, granular, massive, in sands	no, parting {0001}, {10$\bar{1}$1}		6	4.8	black	black
Magnetite, Fe_3O_4	151	C	xl. common, {111}, {110}, {100}, massive, granular	no, often parting {111}	magnetic	6	5.2	black	black, dull to bright
Hematite, Fe_2O_3	161	T	xl. uncommon, scaly, fibrous, granular, earthy etc.	no, parting {0001}, {10$\bar{1}$1}	not magnetic	6	5.3	red, brown, black	reddish brown
Columbite, $(Fe,Mn)(Nb,Ta)_2O_6$	172	O	xl. common, {010} tabular, prismatic	distinct {010}		6-6½	5.2-6.8	black, brownish black	brown to black
Braunite, $Mn^{2+}(Mn^{3+})_6 SiO_{12}$	165	Q	xl. rare, granular, massive	perfect {112}		6-6½	4.8	brownish black, steel-grey	grey, black
Marcasite, FeS_2	118	O	xl. common, {010} tabular, prismatic, radiated	distinct {101}	brittle	6-6½	4.9	light brass-coloured, whitish	greyish black
Pyrite, FeS_2	114	C	xl. common {100}, {210}, {111}, granular, massive	no	brittle	6-6½	5.0	brass-coloured	black, greenish tinge
Sperrylite, $PtAs_2$	119	C	xl. {100}, {111}	indistinct {100}	brittle	6½	10.6	tin-white	black

Table 2. Minerals with non-metallic lustre, arranged according to hardness and density

Mineral	Page	cs	Habit	Cleavage	Brittleness etc.	H	D	Colour	Lustre
Talc, $Mg_3Si_4O_{10}(OH)_2$	337	A, M	xl. rare, foliated, dense	perfect {001}	flexible, inelastic, sectile, greasy	1	2.8	pale green, white, grey	greasy, pearly
Natron, $Na_2CO_3 \cdot 10H_2O$	205	M	crusty, as coatings			1-1½	1.5	white, grey, yellowish	vitreous
Pyrophyllite, $Al_2Si_4O_{10}(OH)_2$	337	A, M	xl. rare, radiated, foliated, dense	perfect {001}	flexible, inelastic, sectile	1-1½	2.8	white, pale yellowish, greenish	greasy, pearly
Aurichalcite, $(Zn,Cu)_5(CO_3)_2(OH)_6$	203	M	Crusty			1-2	4.2	light green, dark green, sky-blue	silky
Sal ammoniac, NH_4Cl	134	C	xl. {211}	no distinct		1½	2.0	colourless, yellowish, brownish	vitreous
Nitratine, $NaNO_3$	206	T	Granular			1½-2	2.2	colourless	vitreous
Vivianite, $Fe_3(PO_4)_2 \cdot 8H_2O$	242	M	xl. common, prismatic, {010}, {100}, as coatings, earthy	perfect {010}	flexible	1½-2	2.7	colourless, darkens to bluish, greenish or black	vitreous
Orpiment, As_2S_3	112	M	xl. uncommon, foliated masses	perfect {010}	flexible, inelastic, sectile	1½-2	3.5	lemon-coloured, brownish yellow, streak pale yellow	resinous
Realgar, AsS	112	M	xl. uncommon, massive, granular	good {010}	sectile	1½-2	3.6	red to orange-yellow, streak orange-yellow	resinous
Sulphur, S	88	O	xl. common, {hkl}, massive or crusty	no distinct	brittle	1½-2½	2.1	sulphur-yellow, greenish, brownish, streak white	greasy to adamantine
Erythrite, $Co_3(AsO_4)_2 \cdot 8H_2O$	243	M	xl. rare, powdered coatings	perfect {010}		1½-2½	3.1	crimson-red, streak lighter red	vitreous
Carnallite, $(K,NH_4)MgCl_3 \cdot 6H_2O$	142	O	xl. rare, granular	no		2	1.6	colourless, milk-white, yellowish, reddish	greasy, dull

Mineral	*Page*	cs	Habit	Cleavage	Brittle-ness etc.	H	D	Colour	Lustre
Melanterite, $FeSO_4 \cdot 7H_2O$	218	M	xl. rare, stalactitic, crusty, as coatings	perfect {001}, {110}		2	1.9	greenish, bluish	vitreous
Gypsum, $CaSO_4 \cdot 2H_2O$	215	M	xl. common, {010}, {120}, {11$\bar{1}$}, granular, fibrous	perfect {010}, also {100}, {011}	flexible, inelastic	2	2.3	colourless, white, grey, etc.	vitreous, pearly, silky
Kaolinite, $Al_2Si_2O_5(OH)_4$	335	A	xl. not seen, clayey, earthy	perfect {001}	greasy, plastic in water	2	2.6	white, reddish, brownish	dull, earthy
Carnotite, $K_2(UO_2)_2(VO_4)_2 \cdot 3H_2O$	248	M	xl. rare, powdery,	perfect {001}	radio-active	2	4-5	light yellow, greenish yellow	earthy, dull
Chlorargyrite, $AgCl$	133	C	xl. rare, wax-like, horny masses	no	sectile	2	5.6	colourless, grey, yellow-ish, darkens, streak shiny	resinous, dull
Borax, $Na_2B_4O_5(OH)_4 \cdot 8H_2O$	206	M	xl. prismatic, {100}, {110}, {010}, {001}	perfect {100}, {110}	brittle	2-2½	1.7	colourless, white, grey, yellowish	vitreous, dull
Epsomite, $MgSO_4 \cdot 7H_2O$	218	O	xl. rare, fibrous, crusty, as coatings	perfect {010}, {101}	bitter taste	2-2½	1.7	colourless, white	vitreous, silky
Autunite, $Ca(UO_2)_2(PO_4)_2 \cdot 10H_2O$	247	Q	xl. {001}, in fans, scaly, crusty	perfect {001}	radio-active	2-2½	3.1-3.2	lemon-coloured, light green	vitreous, pearly
Proustite, Ag_3AsS_3	125	T	xl. {10$\bar{1}$0}, {*hkil*}, massive	distinct {10$\bar{1}$1}	brittle	2-2½	5.8	scarlet-vermilion, dark-ens, streak vermilion	adaman-tine
Chrysocolla, $(Cu,Al)_2 H_2Si_2O_5(OH)_4 \cdot nH_2O$	351	O?	cryptocrystalline, crusty, botryoidal, fibrous, earthy	no		2-4	1.9-2.4	greenish, bluish, brown, black	wax-like
Kernite, $Na_2B_4O_6(OH)_2 \cdot 3H_2O$	207	M	Large xl., coarse granular	perfect {100}, {001}		2½	1.9	colourless, white	vitreous

Mineral	*Page*	cs	Habit	Cleavage	Brittle-ness etc.	H	D	Colour	Lustre
Sylvite, KCl	*132*	C	xl. common, {100}, granular, compact	perfect {100}	slightly brittle	2½	2.0	colourless, white	vitreous
Ulexite, $NaCaB_5O_6(OH)_6 \cdot 5H_2O$	*207*	A	xl. needle-shaped, fibrous, capillary			2½	2.0	white	silky
Halite, NaCl	*130*	C	xl. common, {100}, granular, compact	perfect {100}		2½	2.2	colourless, white, yel-lowish, red-dish, bluish	vitreous
Chalcanthite, $CuSO_4 \cdot 5H_2O$	*217*	A	xl. rare, stalactitic, crusty, as coatings	no distinct		2½	2.3	blue	vitreous
Brucite, $Mg(OH)_2$	*176*	T	xl. rare, foliated, massive or fibrous	perfect {0001}	flexible, inelastic, sectile	2½	2.4	whitish, pale green, grey, brown, blue	vitreous, wax-like, pearly
Clinochlore, $(Mg,Al)_6(Si,Al)_4O_{10}(OH)_8$	*345*	A, M	xl. rare, foliated, fine scaly, granu-lar	perfect {001}	flexible, inelastic	2½	2.7-2.9	green, yellowish, brown, violet	vitreous, often dull
Muscovite, $KAl_2(Si_3Al)O_{10}(OH,F)_2$	*339*	M	xl. tabular, conic, lamellar, foliated, scaly	perfect {001}	flexible, elastic	2½	2.8	colourless, yellowish, greenish, brownish	vitreous, pearly
Cryolite, Na_3AlF_6	*138*	M	xl. rare, cube-like, massive, granular	parting {110}, {001}		2½	3.0	colourless, white, brownish, reddish, nearly black	greasy vitreous
Lampro-phyllite, $Sr_2Na_3Ti_3(Si_2O_7)_2(OH)_4$	*281*	M	xl. tabular, foli-ated	perfect {100}		2½	3.3	golden yellow, brownish	strong vitreous, submetal-lic
Uranophane, $Ca(UO_2)_2(SiO_3OH)_2 \cdot 5H_2O$	*273*	M	xl. needle-shaped in radiated spheres, felted	good {100}		2½	3.9	yellow, orange-yellow	vitreous
Pyrargyrite, Ag_3SbS_3	*124*	T	xl. {10$\bar{1}$0}, {$hkil$}, complex, massive, granular	distinct {10$\bar{1}$1}	brittle	2½	5.8	deep red, darkens, streak red	adaman-tine

Mineral	Page	cs	Habit	Cleavage	Brittle-ness etc.	H	D	Colour	Lustre
Cinnabar, HgS	*111*	T	xl. common, rhombohedral, tabular, granular, massive	perfect {10$\bar{1}$0}		2½	8.1	vermilion-red, brownish, streak red	adamant-ine, dull
Biotite, K(Mg,Fe)$_3$ (Si$_3$Al)O$_{10}$ (OH,F)$_2$	*341*	M	xl. tabular, foliated, scaly	perfect {001}	flexible, elastic	2½-3	2.8	dark brown, green, black	vitreous, submetal-lic
Phlogopite, K(Mg,Fe)$_3$ (Si$_3$Al)O$_{10}$ (F,OH)$_2$	*340*	M	xl. tabular, conic, lamellar, foliated	perfect {001}	flexible, elastic	2½-3	2.9	pale yellow, brown	vitreous, pearly
Crocoite, PbCrO$_4$	*223*	M	xl. prismatic, striated, colum-nar, granular	distinct {110}		2½-3	6.0	red, orange-red, streak orange-yellow	strong vitreous
Phosgenite, Pb$_2$CO$_3$Cl$_2$	*204*	Q	xl., prismatic, {110}, {001}, {111}	perfect {110}, {001}		2½-3	6.1	yellowish white, yellow-brown, grey	resinous
Anglesite, PbSO$_4$	*213*	O	xl. uncommon, granular, globular, dense	{001}, {210}		2½-3	6.3	colourless, white, grey, yellowish, greenish	resinous
Gibbsite, Al(OH)$_3$	*177*	M	xl. rare, radiated, crusty, earthy	perfect {001}		2½-3½	2.4	white, grey	vitreous, pearly
Calcite, CaCO$_3$	*185*	T	xl. common, {10$\bar{1}$0}, {10$\bar{1}$1}, {*hkil*}, granular, stalactitic etc.	perfect {10$\bar{1}$1}	high bi-refrin-gence	3	2.7	colourless, white etc.	vitreous
Lepidolite, K(Li,Al)$_3$ (Si,Al)$_4$O$_{10}$ (F,OH)$_2$	*342*	M	xl. uncommon, scaly	perfect {001}	flexible, elastic	3	2.8	lilac, pink, col-ourless, grey, yellowish	pearly
Enargite, Cu$_3$AsS$_4$	*127*	O	xl. common, {001}, tabular, granular, colum-nar	perfect {110}, also {100}, {010}	brittle	3	4.5	steel-grey, black, streak black	sub-adamant-ine, dull
Wulfenite, PbMoO$_4$	*226*	Q	xl. common, tabu-lar, {001}, granular	distinct {101}		3	6.8	yellow, orange-yellow, orange-red, white, grey	adamant-ine, resinous

Mineral	*Page*	cs	Habit	Cleavage	Brittle-ness etc.	H	D	Colour	Lustre
Vanadinite, $Pb_5(VO_4)_3Cl$	241	H	xl. common, prismatic, {10$\bar{1}$0}, {0001}, globular, crusty	no		3	6.9	red, orange-red, reddish brown	sub-adamant-ine
Polyhalite, $K_2Ca_2Mg(SO_4)_4$ (2H_2O	220	A	xl. rare, massive, fibrous	perfect {10$\bar{1}$}		3-3½	2.8	colourless, white	vitreous
Atacamite, $Cu_2Cl(OH)_3$	143	O	xl. prismatic, {$hk0$}, columnar, granular, fibrous	perfect {010}		3-3½	3.8	bright to deep green, streak apple-green	vitreous
Celestine, $SrSO_4$	212	O	xl. tabular, {001}, prismatic, granular	perfect {001}, good {210}		3-3½	4.0	bluish, colourless, greenish, reddish	vitreous
Witherite, $BaCO_3$	199	O	xl. uncommon, globular, granular, fibrous	distinct {010}		3-3½	4.3	white, grey	vitreous
Baryte, $BaSO_4$	211	O	xl. common, prismatic, tabular, {001}, {210}, granular	perfect {001}, {210}		3-3½	4.5	colourless, white, light blue or green, yellowish	vitreous
Cerussite, $PbCO_3$	199	O	xl. common, {010}, {111}, {021}, reticular, granular	distinct {110}		3-3½	6.6	colourless, white, grey	adamant-ine
Laumontite, $Ca(Al_2Si_4)O_{12}\cdot$ 4H_2O	388	M	xl. prismatic, massive	perfect {010}, {110}	often chalk-like	3-4	2.3	white, colourless, pink	vitreous
Stilpnomelane, $K(Fe,Al)_8(Si,Al)_{12}$ $(O,OH)_{36}\cdot nH_2O$	350	A	xl. rare, foliated	perfect {001}	brittle	3-4	2.9	dark green, black, golden or reddish brown	vitreous, sub-metallic
Antigorite, $Mg_3Si_2O_5(OH)_4$	333	M, O	xl. not seen, dense	not seen		3-5	2.6	greenish, mottled, yellow, brown, red, grey	greasy
Chrysotile, $Mg_3Si_2O_5(OH)_4$	333	M, O	xl. not seen, fibrous	not seen		3-5	2.6	greenish, mottled, yellow, brown, red, grey	silky

Mineral	Page	cs	Habit	Cleavage	Brittle-ness etc.	H	D	Colour	Lustre
Kainite, $KMg(SO_4)$ $Cl·3H_2O$	222	M	xl. rare, granular	perfect {001}		$3\frac{1}{2}$	2.1	colourless	vitreous
Kieserite, $MgSO_4·H_2O$	221	M	xl. rare, granular	perfect {110}, {111}		$3\frac{1}{2}$	2.6	colourless, grey, yellowish	vitreous
Anhydrite, $CaSO_4$	214	O	xl. uncommon, granular, massive, fibrous	perfect {010}, also {100}, {001}		$3\frac{1}{2}$	3.0	colourless, grey, bluish	vitreous, pearly
Astrophyllite, $(K,Na)_3(Fe,Mn)_7$ $Ti_2Si_8(O,OH)_{31}$	325	A	xl. tabular, needle-shaped	perfect {001}	folia brittle	$3\frac{1}{2}$	3.4	brass-coloured, dark brown	vitreous, submetallic
Adamite, $Zn_2AsO_4(OH)$	233	O	xl. many faces, in fan-shaped rosettes	good {101}		$3\frac{1}{2}$	4.4	honey-yellow, brown, pale green, white, colourless	vitreous
Descloizite, $PbZnVO_4$ (OH)	235	O	xl. prismatic, {110}, {111}, fibrous, granular	no		$3\frac{1}{2}$	6.2	brown, brownish black, reddish brown	greasy
Heulandite, $(Na,K)Ca_4$ $(Al_9Si_{27})O_{72}·$ $24H_2O$	389	M	xl. common, tabular {010}	perfect {010}		$3\frac{1}{2}$-4	2.2	colourless, white, yellowish, reddish	vitreous
Stilbite, $NaCa_4$ $(Al_9Si_{27})O_{72}·$ $30H_2O$	390	M	xl. common, sheaf-like	perfect {010}		$3\frac{1}{2}$-4	2.2	colourless, white, grey, yellowish, brownish	vitreous, pearly
Wavellite, $Al_3(PO_4)_2$ $(OH,F)_3·$ $5H_2O$	244	O	xl. rare, radiated spheres	good {110}, {101}		$3\frac{1}{2}$-4	2.4	colourless, grey, yellowish, greenish	vitreous
Alunite, $KAl_3(SO_4)_2$ $(OH)_6$	221	T	xl. uncommon, dense, granular, earthy	distinct {0001}		$3\frac{1}{2}$-4	2.8	white, yellowish, reddish	vitreous
Dolomite, $CaMg(CO_3)_2$	192	T	xl. simple {10$\bar{1}$1}, granular	perfect {10$\bar{1}$1}		$3\frac{1}{2}$-4	2.9	white, grey, greenish, brownish, pinkish	vitreous

Mineral	Page	cs	Habit	Cleavage	Brittle-ness etc.	H	D	Colour	Lustre
Aragonite, $CaCO_3$	196	O	xl. common, prismatic, twinned, columnar, crusty, pisolitic	distinct {010}		3½-4	2.9	colourless, white, grey, yellowish, bluish, greenish	vitreous
Rhodochrosite, $MnCO_3$	191	T	xl. {10$\bar{1}$1}, granular, botryoidal	perfect {10$\bar{1}$1}		3½-4	3.6	pink to deep red	vitreous
Azurite, $Cu_3(CO_3)_2(OH)_2$	202	M	xl. common, prismatic, massive, botryoidal	perfect {011}		3½-4	3.8	azure-blue to deep blue, streak light blue	vitreous
Strontianite, $SrCO_3$	198	O	xl. uncommon, columnar, granular, fibrous	good {110}		3½-4	3.8	white, grey, yellowish, greenish	vitreous
Malachite, $Cu_2CO_3(OH)_2$	201	M	xl. rare, massive, botryoidal, stalactitic	perfect {$\bar{2}$01}, distinct {010}		3½-4	4.0	light to dark green, streak light green	vitreous, silky
Sphalerite, ZnS	100	C	xl. common, {111}, {1$\bar{1}$1}, {100}, {110}, massive	perfect {110}		3½-4	4.0	yellowish, brownish to nearly black, streak yellow to brown	adamantine, sub-metallic
Brochantite, $Cu_4SO_4(OH)_6$	218	M	xl. uncommon, crusty, granular	perfect {100}		3½-4	4.0	emerald-green, greenish black	vitreous
Cuprite, Cu_2O	145	C	xl. common, {111}, {100}, {110}, massive, earthy	distinct {111}	brittle	3½-4	6.1	red to nearly black, streak reddish brown	adamantine, sub-metallic
Pyromorphite, $Pb_5(PO_4)_3Cl$	238	H	xl. common {10$\bar{1}$0}, {0001}, {10$\bar{1}$1}, barrel-shaped, globular	indistinct {10$\bar{1}$1}		3½-4	7.0	yellowish, brownish, greenish	resinous
Mimetite, $Pb_5(AsO_4)_3Cl$	239	H	xl. {10$\bar{1}$0}, {0001}, barrel-shaped, globular	indistinct {10$\bar{1}$1}		3½-4	7.3	light yellow, yellowish brown, orange-yellow	resinous
Magnesite, $MgCO_3$	189	T	xl. uncommon, earthy, porcelain-like, granular	perfect {10$\bar{1}$1}		4	3.0	white, grey, yellowish, brownish	vitreous
Margarite, $CaAl_2(Si_2Al_2)O_{10}(OH)_2$	344	M	xl. rare, foliated, scaly	perfect {001}	brittle	4	3.0	white, reddish, pearly-grey	pearly
Fluorite, CaF_2	135	C	xl. common, {100}, {111}, {110}, {hk0}, {hkl}, granular	perfect {111}		4	3.2	light green, blue-green, violet, colourless, yellow, brown etc.	vitreous

Mineral	Page	cs	Habit	Cleavage	Brittle-ness etc.	H	D	Colour	Lustre
Siderite, $FeCO_3$	190	T	xl. {10$\bar{1}$1}, granular	perfect {10$\bar{1}$1}		4	4.0	light to dark brown, reddish brown, greyish, streak white	vitreous
Manganite, $MnO(OH)$	179	M	xl. uncommon, prismatic, columnar, fibrous	perfect {010}, also {110}, {001}		4	4.3	steel-grey to black, streak dark brown to black	sub-metallic
Zincite, ZnO	148	H	xl. rare, granular, foliated	perfect {10$\bar{1}$0}, parting {0001}		4	5.7	red to orange-yellow, streak orange-yellow	resinous
Colemanite, $CaB_3O_4(OH)_3 \cdot H_2O$	207	M	xl. prismatic, granular	perfect {010}, {001}		4-4½	2.4	colourless, white, pale yellowish	vitreous
Pyrosmalite, $(Fe,Mn)_8Si_6O_{15}(OH,Cl)_{10}$	349	T	xl. tabular, columnar, dense, fine granular	perfect {0001}		4-4½	3.1	brownish, olive-green	greasy, sub-metallic
Smithsonite, $ZnCO_3$	192	T	xl. rare, reniform, crusts, stalactitic	perfect {10$\bar{1}$1}		4-4½	4.4	dirty brown, blue, green, yellow, pink, white, streak white	strong vitreous
Chabazite, $Ca(Al_2Si_4)O_{12} \cdot 6H_2O$	392	T	xl. common, cube-like rhombohedra	indistinct {10$\bar{1}$1}		4-5	2.1	colourless, white, yellowish, pink	vitreous
Xenotime, YPO_4	229	Q	xl. as zircon, tabular, {001}	{100}		4-5	4.5-5.1	yellowish, brownish, greyish	greasy
Gmelinite, $Na_4(Al_4Si_8) \cdot O_{24} \cdot 11H_2O$	394	H	xl. simple, tabular	good {10$\bar{1}$0}		4½	2.1	colourless, white, yellowish, reddish	vitreous
Phillipsite, $K(Ca_{0.5}.Na)_2(Si_5Al_3) \cdot O_{16} \cdot 6H_2O$	391	M	xl. twinned	distinct {010}, {100}		4½	2.2	colourless, white, pale yellowish	vitreous
Harmotome, $BaAl_2Si_6O_{16} \cdot 6H_2O$	391	M	xl. twinned	distinct {010}		4½	2.5	colourless, white	vitreous

Mineral	*Page*	cs	Habit	Cleavage	Brittle-ness etc.	H	D	Colour	Lustre
Variscite, $AlPO_4 \cdot 2H_2O$	*241*	O	cryptocrystalline			4½	2.5	apple-green	wax-like
Triphylite, $LiFePO_4$	*229*	O	xl. rare, coarse granular	good {001}, {010}		4½-5	3.6	bluish to greenish grey	vitreous, resinous
Pseudomala-chite, $Cu_5(PO_4)_2$ $(OH)_4$	*233*	M	xl.rare, fibrous, botryoidal, banded	no distinct		4½-5	4.3	emerald-green, green-ish black	vitreous
Scheelite, $CaWO_4$	*224*	Q	xl. common, {101}, {112}, granular	distinct {101}		4½-5	6.1	white, yellow, brown	greasy, resinous
Kyanite, Al_2SiO_5	*265*	A	xl. common, {100}, {010}, foli-ated	perfect {100}		4½ & 6½	3.6	bluish, whit-ish, greyish, greenish	vitreous, pearly
Apophyllite, $KCa_4Si_8O_{20}$ $(F,OH) \cdot 8H_2O$	*348*	Q	xl. common, pris-matic, striated	perfect {001}		5	2.4	colourless, white, grey, pale greenish or yellowish	vitreous
Wollastonite, $CaSiO_3$	*317*	A	xl. uncommon, radiated, colum-nar, foliated, fi-brous	perfect {100}, {001}, {$\bar{1}$01}		5	2.9	colourless, white, grey	vitreous, silky
Pectolite, $NaCa_2Si_3O_8$ (OH)	*319*	A	xl. rare, needle-shaped, radiated	perfect {100}, {001}		5	2.9	colourless, white, grey	vitreous, silky
Datolite, $CaBSiO_4(OH)$	*271*	M	xl. prismatic, granular, porcelain-like	no distinct		5	3.0	colourless, white, pale greenish	vitreous
Apatite, $Ca_5(PO_4)_3F$	*236*	H	xl. common {0001}, {10$\bar{1}$0}, {10$\bar{1}$1}, granular	no distinct		5	3.2	yellowish green, greyish or bluish green, brown-ish, colour-less, dirty	vitreous, greasy
Dioptase, $CuSiO_3 \cdot H_2O$	*291*	T	xl. common, prism, rhombo-hedron	good {10$\bar{1}$1}		5	3.3	emerald-green	vitreous

Mineral	Page	cs	Habit	Cleavage	Brittle-ness etc.	H	D	Colour	Lustre
Hemimorphite, $Zn_4Si_2O_7(OH)_2 \cdot H_2O$	276	O	xl. common, hemihedral, fan-shaped, stalactitic, as coatings	perfect {110}	pyro- & piezo-electric	5	3.4	white, bluish, greenish, yellowish, brownish	vitreous
Analcime, $Na(AlSi_2)O_6 \cdot H_2O$	384	C	xl. {211}, granular	no		5-5½	2.3	colourless, white, grey-ish, yellowish, reddish	vitreous
Natrolite, $Na_2(Si_3Al_2)O_{10} \cdot 2H_2O$	384	O	xl. common, pris-matic, needle-shaped, fibrous, granular	perfect {110}		5-5½	2.3	colourless, white, yellowish, reddish	vitreous
Thomsonite, $NaCa_2(Al_5Si_5)O_{20} \cdot 6H_2O$	386	O	xl. uncommon, prismatic, fan-shaped, spherical	perfect {010}		5-5½	2.3	white, grey, yellow, red	vitreous
Goethite, $FeO(OH)$	180	O	xl. rare, prismatic, {010}, massive, fibrous, earthy	perfect {010}, {100}		5-5½	3.3-4.4	dark brown, yellowish brown, red-dish brown, streak yellow-ish brown	adamant-ine or submetal-lic, dull, silky
Titanite, $CaTiO(SiO_4)$	269	M	xl. common, wedge-shaped, granular, lamellar	distinct {110}		5-5½	3.4-3.6	brownish, yel-lowish, green-ish, greyish, black	weak adamant-ine
Pyrochlore, $(Na,Ca)_2Nb_2(O,OH,F)_7$	166	C	xl. {111}, granular, massive	good {111}		5-5½	4.5	brown, black, yellowish or reddish tinge	vitreous
Monazite, $(Ce,La,Nd)PO_4$	229	M	xl. uncommon, granular, as sand	good {100}, {010}	radio-active	5-5½	4.6-5.4	yellowish, reddish brown	resinous
Opal, $SiO_2 \cdot nH_2O$	363	-	massive, crusty, botryoidal, stalac-titic	con-choidal fracture		5-6	2.0-2.2	colourless, white, grey, pale yellow, brown etc.	vitreous, wax-like, dull
Lazurite, $Na_3Ca(Si_3Al_3)O_{12}S$	379	C	xl. rare, massive	indistinct {110}		5-6	2.4	azure-blue, greenish blue, streak light blue	vitreous

Mineral	*Page*	cs	Habit	Cleavage	Brittle-ness etc.	H	D	Colour	Lustre
Cancrinite, $(Na,Ca)_8$ $(Si_6Al_6)O_{24}$ $(CO_3)_2 \cdot 2H_2O$	*381*	H	xl. uncommon prismatic, massive	$\{10\bar{1}0\}$		5-6	2.5	yellowish, white, reddish	vitreous
Scapolite, $(Ca,Na)_4$ $(Si,Al)_{12}O_{24}$ $(CO_3.SO_4.Cl)$	*381*	Q	xl. common {100}, {110}, {111}, faces uneven, granular	distinct {100}, {110}		5-6	2.5-2.7	white, grey, greenish, yellowish, bluish, reddish	vitreous
Turquoise, $CuAl_6(PO_4)_4$ $(OH)_8 \cdot 4H_2O$	*246*	A	xl. very rare, cryptocrystalline, dense, crusty	no	brittle	5-6	2.6-2.8	blue, bluish green, green	wax-like, dull
Catapleiite, $Na_2ZrSi_3O_9 \cdot$ $2H_2O$	*288*	M	xl. {001}, tabular, rosette-like aggregates			5-6	2.8	colourless, honey-yellow, brownish, bluish, greenish	vitreous
Lazulite, $(Mg,Fe)Al_2$ $(PO_4)_2(OH)_2$	*233*	M	xl. pyramidal, {111}, {$\bar{1}$11}, granular	no marked		5-6	3.1	sky-blue, bluish green or white	vitreous
Enstatite, $(Mg,Fe)SiO_3$	*306*	O	xl. rare, granular, fibrous	distinct {210}		5-6	3.2	grey, green, brown, nearly black	vitreous, submetal-lic
Aeschynite, (Ce,Ca,Fe) $(Ti,Nb)_2$ $(O,OH)_6$	*173*	O	xl. uncommon, prismatic, massive		metamict	5-6	4.2-5.3	dark brown, streak brown to black	metallic
Romanè-chite, $(Ba,H_2O)_2$ Mn_5O_{10}	*171*	O	crusty, earthy			5-6	4.7	black, brownish black, streak black or brownish black	metallic
Uraninite, UO_2	*175*	C	xl. {100}, {111}, massive, banded	no	radio-active	5-6	6.5-11	black, brownish black, streak brownish black	pitchy, submetal-lic, dull
Fergusonite, $YNbO_4$	*174*	Q	xl. prismatic, granular		metamict	5-6½	4.2-5.7	black, brownish black	sub-metallic

Mineral	Page	cs	Habit	Cleavage	Brittle-ness etc.	H	D	Colour	Lustre
Eudialyte, $Na_{15}Ca_6Fe_3Zr_3Si$ $(Si_{25}O_{73})(O,OH,$ $H_2O)_3(Cl,OH)_2$	297	T	xl. uncommon, granular	no		5½	2.9	red, brown, yellowish	vitreous
Brazilianite, $NaAl_3(PO_4)_2$ $(OH)_4$	235	M	xl. prismatic	good {010}		5½	3.0	yellowish, greenish yellow	vitreous
Melilite, $(Ca,Na)_2(Al,Mg)$ $(Si,Al)_2O_7$	275	Q	xl. uncommon, granular	distinct {001}, {110}		5½	3.0	yellowish, brownish, greyish, colourless	vitreous
Aenigmatite, $Na_2(Fe^{2+})_5$ $TiSi_6O_{20}$	325	A	xl. prismatic	perfect {100}, {010}		5½	3.8	dull black, blownish black, streak reddish brown	vitreous, greasy
Willemite, Zn_2SiO_4	253	T	xl. rare, massive, granular	good {0001}		5½	4	whitish, pale yellow, green, brown	vitreous, greasy
Perovskite, $CaTiO_3$	164	O	xl. cube-like, granular	no distinct		5½	4.0	black, brown, yellow, streak grey or white	metallic or subad-amantine
Amblygonite, $(Li,Na)AlPO_4$ (F,OH)	231	A	xl. uncommon, coarse granular	perfect {100}, also {110}, {0$\bar{1}$1}		5½-6	3.0	whitish, pale yellowish, greenish	vitreous, greasy
Tremolite, $Ca_2(Mg,Fe)_5$ $Si_8O_{22}(OH)_2$	307	M	xl. prismatic, ra-diated, colum-nar, fibrous	perfect {110}, 124 & 56° in between	tough when dense	5½-6	3.0-3.4	white, grey, light to dark green, nearly black	vitreous
Hornblende, e.g. $Ca_2(Fe^{2+},Mg)_4$ $(Al,Fe^{3+})(Si_7Al)$ $O_{22}(OH,F)_2$	309	M	xl. prismatic, ra-diated, colum-nar	perfect {110}, 124 & 56° in between		5½-6	3.0-3.4	dark green to black, brownish	vitreous
Arfvedsonite, $Na_3(Fe^{2+},Mg)_4$ $Fe^{3+}Si_8O_{22}(OH)_2$	313	M	xl. prismatic, columnar, granular	perfect {110}		5½-6	3.4	bluish black to black, streak bluish grey	vitreous

413

Mineral	*Page*	cs	Habit	Cleavage	Brittle-ness etc.	H	D	Colour	Lustre
Allanite, Ca(Ce,La) (Al,Fe)$_3$(Si$_2$O$_7$) (SiO$_4$)(O,OH)$_2$	285	M	xl. uncommon, granular, massive	indistinct	meta-mict, radio-active	5½-6	3.5-4.2	pitchy black, dark brown, streak dark brown	vitreous, greasy, sub-metallic
Periclase, MgO	146	C	xl. rare, granular	perfect {100}		5½-6	3.6	colourless, yellowish brown or green	vitreous
Anatase, TiO$_2$	171	Q	xl. bipyramidal, {101}, {100}, tabular	perfect {001}, {101}		5½-6	3.9	bluish black, yellow, brown etc., streak whitish	adamant-ine, sub-metallic
Ilvaite, CaFe^{3+}(Fe^{2+})$_2$O (Si$_2$O$_7$)(OH)	279	M	xl. common, prismatic, striated, massive, radiated	distinct {010}, {001}		5½-6	4.0	black, brownish black, streak nearly black	vitreous, submetal-lic
Brookite, TiO$_2$	172	O	xl. {010} tabular, {120} prismatic	indistinct {120}		5½-6	4.1	yellowish brown, reddish brown, black	subadam-antine
Hausmannite, Mn$_3$O$_4$	155	Q	xl. uncommon, massive, granular	perfect {001}		5½-6	4.8	brown to black, streak brownish	submetal-lic
Sodalite, Na$_4$(Si$_3$Al$_3$) O$_{12}$Cl	376	C	xl. rare, {110}, granular, massive	indistinct {110}		6	2.3	bluish, greenish, reddish, white, grey	vitreous
Milarite, (K,Na)Ca$_2$ (Be,Al)$_3$Si$_{12}$O$_{30}$· H$_2$O	296	H	xl. prismatic			6	2.5	colourless, light green, yellow	vitreous
Leucite, K(AlSi$_2$)O$_6$	376	Q, C	xl. common as cubic {211}	no		6	2.5	white, grey, colourless	vitreous, dull
Nepheline, (Na,K)AlSiO$_4$	375	H	xl. rare, granular	indistinct {10$\bar{1}$0}		6	2.6	colourless, white, grey	greasy
Sanidine, KAlSi$_3$O$_8$	366	M	xl. {010} tabular	perfect {001}, {010}		6	2.6	colourless, white, grey, bluish milky tinge	vitreous
Orthoclase, KAlSi$_3$O$_8$	367	M	xl. common, prismatic, {010}, {110}, {001}, granular	perfect {001}, {010}		6	2.6	weakly flesh-red, white, grey, yellowish, greenish	vitreous

Mineral	*Page*	cs	Habit	Cleavage	Brittle-ness etc.	H	D	Colour	Lustre
Microcline, $KAlSi_3O_8$	*369*	A	xl. common, prismatic, {010}, {110}, {001}, granular	perfect {001}, {010}		6	2.6	white, pale yellow, red-dish, greenish	vitreous
Plagioclase, $(Na,Ca)(Si,Al)_4O_8$	*371*	A	xl. tabular after {010}, granular	perfect {001}, {010}	twin-striations on {001}	6	2.6-2.8	white, grey, colourless, greenish, reddish	vitreous
Anthophyllite, $(Mg,Fe)_7Si_8O_{22}(OH)_2$	*315*	O	xl. rare, lamellar, fibrous	perfect {210}		6	2.8-3.3	brown, yellowish, greyish, greenish	vitreous, silky
Glaucophane, $Na_2(Mg,Fe)_3Al_2Si_8O_{22}(OH)_2$	*310*	M	xl. prismatic, needle-shaped, radiated, granular	distinct {110}		6	3.0-3.2	bluish grey, blue to nearly black	vitreous
Lawsonite, $CaAl_2Si_2O_7(OH)_2 \cdot H_2O$	*278*	O	granular	perfect {001}, {100}, also {110}		6	3.1	white, light bluish or greyish	vitreous
Cummingtonite, $(Mg,Fe,Mn)_7Si_8O_{22}(OH)_2$	*315*	M	xl. rare, needle-shaped, fibrous, radiated	perfect {110}		6	3.2-3.6	light to dark brown, greenish	vitreous, silky
Diopside, $CaMgSi_2O_6$	*300*	M	xl. prismatic, granular, massive, columnar	distinct {110}, 87 & 93° in between		6	3.3	white, light to dark green	vitreous
Augite, $(Ca,Na)(Mg,Fe,Al)(Si,Al)_2O_6$	*302*	M	xl. prismatic, granular, massive	distinct {110}, 87 & 93° in between		6	3.3	black, deep green	vitreous
Bustamite, $CaMnSi_2O_6$	*318*	A	xl. uncommon, compact, fibrous	perfect {100}, also {110}, {1$\bar{1}$0}		6	3.3	light red, brownish red	vitreous
Helvite, $Be_3Mn_4(SiO_4)_3S$	*381*	C	xl. {111}, {1$\bar{1}$1}, massive	distinct {111}		6	3.3	yellow	vitreous

Mineral	*Page*	cs	Habit	Cleavage	Brittle-ness etc.	H	D	Colour	Lustre
Riebeckite, $Na_2(Fe^{2+},Mg)_3$ $(Fe^{3+})_2Si_8O_{22}$ $(OH,F)_2$	*313*	M	xl. uncommon, radiated, felted, fibrous	perfect {110}		6	3.4	blue, bluish black	vitreous, silky
Babingtonite, $Ca_2(Fe^{2+},Mn)$ $Fe3_1Si_5O_{14}$ (OH)	*322*	A	xl. prismatic	perfect {110}, {1$\bar{1}$0}		6	3.4	dark green, black	strong vitreous
Aegirine, $NaFe^{3+}Si_2O_6$	*303*	M	xl. prismatic, needle-shaped, fibrous, granular	distinct {110}		6	3.5	dark green, greenish black, brownish black, streak yellowish green	vitreous
Rhodonite, $(Mn,Ca)_5$ Si_5O_{15}	*320*	A	xl. tabular, massive, granular	perfect {110}, {1$\bar{1}$0}		6	3.5-3.7	pink, brownish red	vitreous
Prehnite, $Ca_2Al(Si,Al)_4$ $O_{10}(OH)_2$	*348*	O	xl. rare, stalactitic, botryoidal, massive	distinct {001}		6-6½	2.9	light to darker green, white, grey	vitreous
Chondrodite, $(Mg,Fe,Ti)_5$ $(SiO_4)_2$ $(F,OH,O)_2$	*256*	M	xl. common, many forms, granular	no distinct		6-6½	3.2	yellow, orange, brownish red	vitreous
Rutile, TiO_2	*166*	Q	xl. common, {100}, {110}, {101}, {111}, granular	distinct {110}, {100}		6-6½	4.2	reddish brown, golden brown, red, black, streak light brown	adamantine, sub-metallic
Braunite, $Mn^{2+}(Mn^{3+})_6$ SiO_{12}	*165*	Q	xl. rare, granular, massive	perfect {112}		6-6½	4.8	brownish black, steel-grey, streak grey to black	metallic
Columbite, (Fe,Mn) $(Nb,Ta)_2O_6$	*172*	O	xl. common, {010} tabular, prismatic	distinct {010}		6-6½	5.2-6.8	black, brownish black, streak brown to black	metallic
Sillimanite, Al_2SiO_5	*264*	O	xl. rare, fibrous, columnar	perfect {010}		6-7	3.2	greyish, yellowish, light brown or green	vitreous, silky

Mineral	Page	cs	Habit	Cleavage	Brittle-ness etc.	H	D	Colour	Lustre
Epidote, $Ca_2FeAl_2(Si_2O_7)$ $(SiO_4)(O,OH)_2$	282	M	xl. elongated after b axis, striated, granular, fibrous	good {001}, {100}		6-7	3.4	yellowish green, dark green, brown-ish green, nearly black	vitreous
Clinozoisite, $Ca_2Al_3(Si_2O_7)$ $(SiO_4)(O,OH)_2$	284	M	xl. elongated after b axis, striated, granular, fibrous	good {001}, {100}		6-7	3.4	yellowish green, dark green, brownish	vitreous
Cassiterite, SnO_2	167	Q	xl. common, {100}, {110}, {101}, {111}, granular	indistinct {100}		6-7	7.0	reddish brown, brownish black, streak white, light yellow	adam-antine to metallic
Cristobalite, SiO_2	363	Q	xl. small, octahedral, spherical	no		6½	2.3	white	vitreous
Petalite, $LiAlSi_4O_{10}$	350	M	xl. rare, granular	perfect {001}		6½	2.4	colourless, white, grey, reddish, greenish	vitreous
Ussingite, $Na_2AlSi_3O_8$ (OH)	382	A	granular	distinct {001}, {110}		6½	2.5	reddish violet, white	vitreous
Kornerupine, $Mg_4Al_6(Si,Al,B)_5$ $O_{21}(OH)$	273	O	xl. prismatic, columnar, radiated	{110}		6½	3.3	colourless, yellowish, greenish, brownish	vitreous
Jadeite, $Na(Al,Fe)Si_2O_6$	304	M	xl. rare, micro-crystalline dense	{110}, rarely seen	tough	6½	3.3	light green, dark green, whitish, brownish	vitreous
Vesuvianite, $(Ca,Na)_{19}$ $(Al,Mg,Fe)_{13}$ $(SiO_4)_{10}(Si_2O_7)_4$ $(OH,F,O)_{10}$	286	Q	xl. common, {100}, {101}, {110}, granular, radiated	no distinct		6½	3.4	green, brown, yellow, blue	vitreous, resinous
Chloritoid, (Fe,Mg,Mn) $Al_2SiO_5(OH)_2$	270	M, A	xl. uncommon, scaly, foliated	good {001}		6½	3.6	greenish, black	vitreous

Mineral	*Page*	cs	Habit	Cleavage	Brittle-ness etc.	H	D	Colour	Lustre
Benitoite, $BaTiSi_3O_9$	*288*	H	xl. common, tri-gonal bipyramid	no distinct		$6\frac{1}{2}$	3.7	light to dark blue	vitreous
Gadolinite, $Be_2FeY_2Si_2O_{10}$	*271*	M	xl. rare, dense masses	no	metamict	$6\frac{1}{2}$	4-4.7	brown to black, streak greenish grey	greasy
Spodumene, $LiAlSi_2O_6$	*304*	M	xl. common, pris-matic, striated	perfect {110}	parting {100}	$6\frac{1}{2}$-7	3.2	white, grey, pink, green, yellow	vitreous
Axinite, $Ca_2(Mn,Fe,Mg)Al_2BSi_4O_{15}(OH)$	*277*	A	xl. common, axe-head-shaped, massive, granular	one good direction		$6\frac{1}{2}$-7	3.3	brown, violet, yellowish, greenish, greyish	vitreous
Olivine, $(Mg,Fe)_2SiO_4$	*254*	O	xl. pinacoid, prism, bipyramid, granular	no distinct		$6\frac{1}{2}$-7	3.3-4.4	yellowish green, olive-green, brown, black	vitreous
Diaspore, $AlO(OH)$	*179*	O	xl. rare, pris-matic, tabular, massive, scaly	perfect {010}, {110}		$6\frac{1}{2}$-7	3.4	white, grey, colourless, greenish, brownish, reddish	vitreous, pearly
Tridymite, SiO_2	*363*	O	xl. ≤ 1 mm, tabular	no		7	2.3	colourless, white	vitreous
Quartz, SiO_2	*356*	T	xl. common, pris-matic with rhom-bohedra, striated, granular, micro-crystalline	no, con-choidal fracture	piezo-electric	7	2.65	colourless, grey, white, violet, yellow, brown, pink	vitreous
Dumortierite, $(Al,Mg,Fe)_{27}B_4Si_{12}O_{69}(OH)_3$	*273*	O	xl. prismatic, dense, fibrous			7	3.3	dark blue, vio-let, greyish blue, brownish, reddish	silky
Pyrope, $Mg_3Al_2(SiO_4)_3$	*259*	C	xl. common, {110}, {211}, granular	no		7	3.6	deep red, nearly black	vitreous
Grossular, $Ca_3Al_2(SiO_4)_3$	*259*	C	xl. common, {110}, {211}, granular	no		7	3.6	white, yellow, pink, green, brown	vitreous

Mineral	Page	cs	Habit	Cleavage	Brittle-ness etc.	H	D	Colour	Lustre
Staurolite, $(Fe,Mg)_4Al_{17}(Si,Al)_8O_{45}(OH)_3$	267	M	xl. common, {110}, {001}, {010}, {101}, twins	distinct {010}		7	3.7	light brown, reddish brown, brownish black	vitreous
Andradite, $Ca_3Fe_2(SiO_4)_3$	259	C	xl. common, {110}, {211}, granular	no		7	3.9	yellow, green, brown or black	vitreous
Uvarovite, $Ca_3Cr_2(SiO_4)_3$	259	C	xl. common, {110}, {211}, granular	no		7	3.9	emerald-green	vitreous
Spessartine, $Mn_3Al_2(SiO_4)_3$	259	C	xl. common, {110}, {211}, granular	no		7	4.2	orange, red, brown	vitreous
Almandine, $Fe_3Al_2(SiO_4)_3$	259	C	xl. common, {110}, {211}, granular	no		7	4.3	red, brown	vitreous
Cordierite, $Mg_2Al_4Si_5O_{18}$	291	O	xl. prismatic, granular	indistinct {010}	pleo-chroic	7-7½	2.6	light to dark blue or violet, colourless, grey	vitreous
Danburite, $CaB_2Si_2O_8$	282	O	xl. prismatic			7-7½	3.0	colourless, light yellowish	vitreous, greasy
Boracite, $Mg_3B_7O_{13}Cl$	208	O	xl. {100}, {110}, {111}, {1$\bar{1}$1}, dense globular	no		7-7½	3.0	white to grey, light greenish or bluish	vitreous, dull
Tourmaline, e.g. $Na(Fe,Mg)_3Al_6(BO_3)_3Si_6O_{18}(OH)_4$	293	T	xl. common, prismatic, striated, hemi-hedral, mas-sive, columnar, radiated	no distinct	pyro- & piezo-electric	7-7½	3.0-3.2	black, brown, green, pink, yellow, colour-less, blue, multicoloured	vitreous
Euclase, $BeAlSiO_4(OH)$	263	M	xl. common, prismatic	perfect {010}		7½	3.0	colourless, light green or blue	vitreous
Andalusite, Al_2SiO_5	264	O	xl. prismatic	good {110}	altered surface	7½	3.2	greyish, greenish, brownish, red-dish, dirty	vitreous

Mineral	*Page*	cs	Habit	Cleavage	Brittle-ness etc.	H	D	Colour	Lustre
Sapphirine, $Mg_7Al_{18}Si_3O_{40}$	*326*	M, A	xl. uncommon tabular, granular	indistinct		7½	3.5	sapphire-blue, greenish	vitreous
Zircon, $ZrSiO_4$	*260*	Q	xl. common, prism, bipyramid	poor {100}	high bi-refrin-gence	7½	4.7	brownish, brownish red, yellow, colour-less, blue	adaman-tine
Beryl, $Be_3Al_2Si_6O_{18}$	*289*	H	xl. common, {10$\bar{1}$0}, {0001}, striated, grooved, columnar	indistinct {0001}		7½-8	2.7	pale green, blue, colour-less, yellow, dark green, pink, red	vitreous
Phenakite, Be_2SiO_4	*253*	T	xl. rhombohedral, prismatic	indistinct prismatic		7½-8	3.0	colourless, whitish	vitreous
Spinel, $MgAl_2O_4$	*150*	C	xl. common, {111}, {110}, {100}, massive	no, in-distinct parting {111}		7½-8	3.6	red, blue, green, brown, colourless, black	vitreous
Topaz, Al_2SiO_4 $(F,OH)_2$	*268*	O	xl. common, pris-matic, striated, granular	perfect {001}		8	3.5	colourless, yellow, brown, blue, green, pink	vitreous
Chrysoberyl, $BeAl_2O_4$	*156*	O	xl. common, {001}, {110}, {010}, twins	distinct {110}, in-distinct {010}		8½	3.7	green, yel-lowish green, brownish, red	vitreous
Corundum, Al_2O_3	*158*	T	xl. common, bar-rel-shaped with {$hki l$}, {0001}, {11$\bar{2}$0}, granular	no, part-ing {0001}, {10$\bar{1}$1}		9	4.0	grey, blue, yellow, red	vitreous
Diamond, C	*88*	C	xl. common, {111}, {110}	perfect {111}	brittle	10	3.5	colourless, yellowish, etc.	adaman-tine

The Periodic Table of the Elements

Key:

Atomic number	14
Symbol	Si

1a	2a	3b	4b	5b	6b	7b	8			1b	2b	3a	4a	5a	6a	7a	0
1 H																	2 He
3 Li	4 Be											5 B	6 C	7 N	8 O	9 F	10 Ne
11 Na	12 Mg											13 Al	14 Si	15 P	16 S	17 Cl	18 Ar
19 K	20 Ca	21 Sc	22 Ti	23 V	24 Cr	25 Mn	26 Fe	27 Co	28 Ni	29 Cu	30 Zn	31 Ga	32 Ge	33 As	34 Se	35 Br	36 Kr
37 Rb	38 Sr	39 Y	40 Zr	41 Nb	42 Mo	43 Tc	44 Ru	45 Rh	46 Pd	47 Ag	48 Cd	49 In	50 Sn	51 Sb	52 Te	53 I	54 Xe
55 Cs	56 Ba	57* La	72 Hf	73 Ta	74 W	75 Re	76 Os	77 Ir	78 Pt	79 Au	80 Hg	81 Tl	82 Pb	83 Bi	84 Po	85 At	86 Rn
87 Fr	88 Ra	89** Ac															

Transition metals

*Lanthanides

58 Ce	59 Pr	60 Nd	61 Pm	62 Sm	63 Eu	64 Gd	65 Tb	66 Dy	67 Ho	68 Er	69 Tm	70 Yb	71 Lu

**Actinides

90 Th	91 Pa	92 U	93 Np	94 Pu	95 Am	96 Cm	97 Bk	98 Cf	99 Es	100 Fm	101 Md	102 No	103 Lr

Symbols and atomic numbers of selected elements*

Name	Symbol	Atomic number	Name	Symbol	Atomic number
Aluminium	Al	13	Nickel	Ni	28
Antimony	Sb	51	Niobium	Nb	41
Argon	Ar	18	Nitrogen	N	7
Arsenic	As	33	Osmium	Os	76
Barium	Ba	56	Oxygen	O	8
Beryllium	Be	4	Palladium	Pd	46
Bismuth	Bi	83	Phosphorus	P	15
Boron	B	5	Platinum	Pt	78
Bromine	Br	35	Potassium	K	19
Cadmium	Cd	48	Radium	Ra	88
Caesium	Cs	55	Rhodium	Rh	45
Calcium	Ca	20	Rubidium	Rb	37
Carbon	C	6	Scandium	Sc	21
Cerium	Ce	58	Selenium	Se	34
Chlorine	Cl	17	Silicon	Si	14
Chromium	Cr	24	Silver	Ag	47
Cobalt	Co	27	Sodium	Na	11
Copper	Cu	29	Strontium	Sr	38
Fluorine	F	9	Sulphur	S	16
Gold	Au	79	Tantalum	Ta	73
Hafnium	Hf	72	Tellurium	Te	52
Helium	He	2	Thorium	Th	90
Hydrogen	H	1	Tin	Sn	50
Iodine	I	53	Titanium	Ti	22
Iridium	Ir	77	Tungsten	W	74
Iron	Fe	26	Uranium	U	92
Lanthanum	La	57	Vanadium	V	23
Lead	Pb	82	Yttrium	Y	39
Lithium	Li	3	Zinc	Zn	30
Magnesium	Mg	12	Zirconium	Zr	40
Manganese	Mn	25			
Mercury	Hg	80	* Only elements mentioned in this		
Molybdenum	Mo	42	book are included.		

Glossary

This glossary includes a selection of geological terms used in the sections on mineral occurrence in Part II. Mineralogical and crystallographic terms are explained in Part I and can be found by referring to the index.

alkali granite, alkali syenite. A granite or syenite with a particular high content of alkali metals, primarily Na and K, which results in the formation of Na-rich pyroxenes and amphiboles.

alkaline rock. An igneous rock with a high content of Na and K and a relatively low content of Al and Si. Feldspathoids are present in addition to alkali feldspars, together with pyroxenes and amphiboles rich in Na.

Alpine vein. A term for various types of hydrothermal veins, generally those containing especially well-developed crystals.

amphibolite. A mafic (dark-coloured) metamorphic rock consisting predominantly of hornblende and plagioclase.

andesite. A light-coloured, fine-grained volcanic rock consisting of plagioclase (oligoclase or andesine), K-feldspar, a little quartz, and variable amounts of dark minerals. Andesites are abundant in island arcs and at active plate margins, e.g. the South American Andes.

anorthosite. A plutonic rock consisting almost entirely of plagioclase (usually labradorite).

basalt. A volcanic rock consisting of Ca-rich plagioclase, pyroxene, and iron oxides, with or without olivine, containing 40–52% SiO_2. The term is also used generally for any dark, fine-grained rock. Basalts are the most abundant of volcanic rocks and are the equivalents of gabbros. They occur as widespread lava flows in the North Atlantic region, India, and elsewhere, and form the bulk of the ocean floors.

basic rock. An igneous or metamorphic rock that is undersaturated, i.e. one that contains 45–53% of SiO_2 by weight.

bentonite. A clay composed mainly of montmorillonite, formed by the alteration of volcanic ash.

bitumen. A mixture of various natural hydrocarbons such as asphalt.

breccia. A rock composed of angular rock fragments. The angular shape indicates that the fragments have not been transported far from their source.

carbonatite. An igneous rock with significant amounts of carbonate minerals, usually calcite or dolomite. Some carbonatites are of economic importance because of their content of, e.g. Nb minerals.

chalk. A porous fine-grained limestone, white or pale yellowish in colour, composed mainly of the remains of micro-organisms, chiefly foraminifera.

clay. A sedimentary rock composed of grains or particles less than 0.004 mm (4 μm) in diameter. The individual particles are clay minerals or fragments of other minerals or rocks.

complex (igneous). A system of closely associated intrusions of various types of igneous rocks, all of about the same age.

concretion. A rounded nodule of a mineral occurring in a sedimentary rock. A concretion usually has a concentric structure, having formed round a nucleus.

conglomerate. A coarse-grained sedimentary rock composed of rounded fragments more than 2 mm in diameter, cemented together by a finer matrix of, e.g. sand, clay, or chalk.

contact metamorphism. See *metamorphism.*

crust. The outermost part of the earth down to the Moho. The crust under the continents is granitic and is generally about 30-40 km thick; oceanic crust is basaltic and about 7–13 km thick.

differentiation. See *igneous rock.*

diorite. A medium- to coarse-grained plutonic rock of intermediate composition, composed essentially of plagioclase (oligoclase or andesine), hornblende, and minor biotite or pyroxene; there is little or no quartz or K-feldspar.

dolomite. (1) A carbonate mineral, $CaMg(CO_3)_2$; (2) a rock (*dolostone*) consisting predominantly of the mineral dolomite. A dolostone resembles a limestone in appearance.

dunite. An ultrabasic rock, usually dark green or brown in colour, consisting of more than 90% olivine.

dyke (dike). A tabular body of igneous rock intruded into an older rock and cutting across it. Dykes are usually vertical or nearly vertical. The grain size is generally intermediate between those of plutonic and volcanic rocks.

eclogite. A rock formed under high pressure and at high temperature, consisting primarily of reddish garnet (pyrope-rich) and greenish pyroxene (omphasite). At least some eclogites are of metamorphic origin.

enriched copper zone. See *chalcocite* (p. 95).

evaporites. Deposits precipitated from sea water or water in saline lakes. The precipitation results from the evaporation of the water and saturation of the solutions. See also *halite* and *gypsum*, pp. 130 and 215.)

fumarole. An opening at the earth's surface from which gases of volcanic origin escape.

Such gases can include water vapour, carbon dioxide, hydrogen sulphide, etc. A fumarole giving off sulphurous gases is sometimes called a *solfatara.*

gabbro. A dark, medium- to coarse-grained plutonic rock consisting of Ca-rich plagioclase and clinopyroxene, usually with minor olivine. Gabbros are the plutonic equivalents of basalts. See also *norite.*

gangue mineral. A mineral of no commercial value that is present in an ore deposit, as opposed to an ore mineral.

gneiss. A medium- to coarse-grained foliated metamorphic rock formed by the high-grade regional metamorphism of sediments or igneous rocks. The minerals are characteristically arranged in alternating light and dark bands. Typical gneisses contain plagioclase, K-feldspar, and quartz with variable amounts of darker minerals such as micas, amphiboles, or pyroxenes.

granite. A coarse-grained plutonic, oversaturated rock containing quartz (at least 20%), K-feldspar, Na-rich plagioclase (with more K-feldspar than plagioclase), and mica (muscovite or biotite, or both), with minor amounts of darker accessory minerals such as apatite and magnetite; hornblende can also be present. Granites are the plutonic equivalents of rhyolites. They form large parts of the continental shields.

granodiorite. A coarse-grained igneous rock differing from granite by having more plagioclase than K-feldspar.

greenschist. A chlorite-bearing metamorphic rock formed at relatively low temperature and moderate pressure.

hornfels. A fine-grained, horn-like metamorphic rock formed by contact metamorphism.

hydrothermal deposit. A mineral deposit formed by precipitation from hydrothermal solutions (hydrous solutions, generally of magmatic origin). A distinction is made between low-temperature deposits (formed at about $50–150\,°C$), intermediate deposits (formed at about 150–400 °C), and high-temperature deposits (formed at about 400–600 °C). Hydrothermal deposits generally occur as veins or replacement deposits (q.v.). Hydrothermal formations are formed at a later stage of magmatic activity than pegmatites.

igneous complex. A system of closely associated igneous intrusions of various types, all of about the same age.

igneous rock. A rock crystallized from a silicate melt (a *magma*). Igneous rocks are divided into volcanic (or extrusive) rocks, formed at or near the surface of the earth; hypabyssal rocks, formed at shallow depth; and plutonic rocks, formed at greater depth. The distinction between hypabyssal rocks and the other two categories is imprecise, and the term *hypabyssal* is avoided by some geologists. Volcanic rocks crystallize relatively rapidly and are consequently fine-grained. Plutonic rocks crystallize slowly and as a result are coarse-grained. Igneous rocks are classified according to their silica (SiO_2) content into oversaturated ('acid' or silicic), saturated, and undersaturated ('basic') rocks. Ultrabasic rocks are extremely undersaturated rocks. Quartz, K-feldspars, and Na-rich plagioclase are the primary minerals in oversaturated rocks, whereas plagioclase or K-feldspars without quartz predominate in saturated rocks. Quartz is also absent from undersaturated rocks, and feldspars are partly replaced by feldspathoids. Ultrabasic rocks consist largely of darker minerals such as pyroxenes or olivine. Granite and basalt are among the most common igneous rocks.

Some igneous rocks are formed as a result of magmatic differentiation. This process takes place in a magma chamber when material that has already crystallized is removed from the magma, whether by loss of part of the magma or by the deposition of heavier crystals at the bottom of the magma chamber. This affects the composition of the residual magma and the mineral composition of the rocks being formed from the magma.

intermediate rock. An igneous rock containing 52–66% silica (SiO_2).

intrusive rock. An igneous rock that has crystallized from magma emplaced within pre-existing rocks.

kimberlite. A diamond-bearing ultrabasic rock, green to black in colour, consisting largely of olivine and sometimes also phlogopite. Kimberlites commonly have a porphyritic texture and some resemble breccias. They are known from, e.g. the Republic of South Africa, where they occur in carrot-shaped pipes reaching depths of 150 km or more, i.e. down to the earth's mantle.

lava. A silicate melt erupted at the earth's surface.

lignite. A low-grade coal deposit such as brown coal, usually with a distinct woody texture.

limestone. A pale-coloured, sedimentary rock, generally white or yellowish-white, composed predominantly of calcium carbonate minerals, i.e. calcite or aragonite, and sometimes of dolomite. Limestones are formed either by the deposition of organic material or by chemical precipitation.

mantle. The region between the Moho, which marks the boundary between the earth's crust and the mantle, and the earth's outer core, at a depth of approximately 2900 km.

marble. A metamorphic rock formed by the metamorphism of a limestone or a dolomite. Marbles consist primarily of calcite and, more rarely, of dolomite, and are usually white in colour. Commercially, any calcareous rock that takes a polish is called 'marble'.

marl. A sedimentary rock consisting of roughly equal amounts of clay and calcium carbonate, i.e. a calcareous mudstone.

massif. (1) A large structural feature; (2) a large topographical feature; (3) a large igneous intrusion or body of metamorphic rock forming a resistant mass.

metamorphism. The processes by which rocks are altered by changes in temperature, pressure, or the presence of fluids. During metamorphism, the minerals of the original rock recrystallize to form new minerals that are stable under the changed conditions. Four main types of metamorphism are recognized: (1) *contact*, or *thermal, metamorphism* taking place at high temperatures, e.g. by contact with a magma (i.e. a silicate melt); (2) *regional metamorphism*, under increasing temperature and pressure during large-scale geological events such as mountain-building; (3) *cataclastic metamorphism*, produced by physical deformation and fracture with little heating, charactistic of fault zones; (4) *impact metamorphism*, caused by shock waves at sites of meteoric impact. A distinction is made between low-, medium-, and high-grade metamorphic rocks according to the degree of transformation. These various degrees of metamorphism give rise to characteristic mineral associations. Gneisses, mica-schists, marbles, and quartzites are examples of common metamorphic rocks. See also *metasomatism*.

metasomatism. A term used for changes in composition produced by the introduction or removal of chemical components, usually by circulating fluids. Rocks affected by metasomatism are termed *metasomatic rocks*.

Moho. The *Mohorovičić discontinuity*, the seismic discontinuity that marks the boundary between the earth's crust (which is essentially granitic) and the upper mantle (which is peridotitic). The Moho is generally at a depth of 25–90 km below the surface of the continents and 7–10 km below the ocean floors.

nepheline-syenite. A syenite composed essentially of alkali feldspar, nepheline, pyroxene, and amphibole.

norite. A gabbro containing more orthopyroxene than clinopyroxene.

obsidian. A black, completely glassy rock with a composition corresponding to that of a rhyolite.

ore mineral. A valuable mineral in an ore deposit, as opposed to a gangue mineral of no commercial value.

pegmatites. Coarse-grained igneous rocks, usually of granitic composition and occurring as dykes or veins, formed during the final stages of crystallization of igneous intrusions. Volatiles (H_2O, F, B, Cl, etc.) that are present at these stages reduce the viscosity of the fluid and facilitate the intrusion of other rocks. The crystals in pegmatites can be extremely large, weighing up to hundreds of tonnes. Some pegmatites contain minerals rich in Li, Be, B, F, P, Nb, Ta, U, and other elements.

peridotite. A coarse-grained dark-coloured ultrabasic rock with olivine as the predominant mineral.

phonolite. A fine-grained porphyritic volcanic rock corresponding in composition to a nepheline-syenite.

plutonic rock. See *igneous rock*.

pneumatolysis. A process in which the chemistry and mineralogy of a rock are changed by reactions with hot gases derived from a magma during the late stages of crystallization. The gases are rich in F, Cl, S, B, and CO_2 and contribute to the formation of minerals such as topaz, tourmaline, cassiterite, molybdenite, and scheelite. Genetically, pneumatolysis is intermediate between the formation of pegmatites and of hydrothermal veins.

porphyry. An intrusive igneous rock with relatively large mineral grains (phenocrysts) embedded in a finer-grained matrix.

pyroxenite. An ultrabasic rock consisting predominantly of pyroxenes, usually with minor olivine.

quartzite. A metamorphic rock consisting essentially of quartz. Quartzites are usually formed by the metamorphism of sandstones.

replacement deposit. An orebody formed by the hydrothermal mineralization of an existing rock. Replacement is usually a volume-for-volume process, as opposed to a filling of a vein or a similar opening.

rhyolite. A fine-grained, generally pale-coloured volcanic rock with a composition corresponding to that of a granite. Some rhyolites are partly glassy.

rock. An aggregate of grains of one or more minerals forming more-or-less uniform masses on a geological scale. Rocks are broadly classified into igneous, sedimentary, and metamorphic rocks according to their origin.

sandstone. A sedimentary rock consisting of sand grains, cemented together by silica, calcite, iron oxides, or clay minerals, etc. The sand grains, 0.06–2 mm in diameter, are usually quartz but can also be feldspars or other minerals.

schist. A distinctly foliated (*schistose*) metamorphic rock. Schists are generally named after a characteristic mineral, e.g. talc, chlorite, mica, garnet, etc. Some of these minerals are indicative of the grade of metamorphism.

sedimentary rock. A rock formed at the earth's surface by the accumulation of fragments or particles, whether mineral grains, rock fragments, or organic remains. Most sediments are produced by the erosion and transport of existing rocks; others are formed chemically by precipitation from solutions, by the accumulation of organic material, and by the deposition of fragments ejected by volcanoes. Common sediments are clay, sand, sandstone, and gravel.

serpentinite. A rock consisting essentially of serpentine minerals. Serpentinites are formed by the alteration of olivine-rich ultrabasic rocks such as peridotites or dunites.

shale. A consolidated, very fine-grained sedimentary rock that splits easily (i.e. is fissile).

skarn. A contact-metamorphic rock composed of silicate minerals. Skarns are derived from limestones and dolomitic limestones that have been influenced by metasomatism, and are associated with certain iron ores and other ore deposits. Typical skarn minerals include grossular, andradite, diopside, and epidote.

sublimation. Crystallization directly from a gaseous phase, e.g. the formation of sulphur crystals from sulphurous vapours.

syenite. A medium- to coarse-grained plutonic rock consisting primarily of feldspars, of which two-thirds or more are K-feldspars or Na-rich plagioclase, together with amphiboles, micas, or pyroxenes; there is little or no quartz.

trachyte. A volcanic rock corresponding in composition to a syenite.

travertine. A calcareous deposit formed under atmospheric conditions at a thermal spring.

ultrabasic rock. An igneous or metamorphic rock that is extremely undersaturated, i.e. with a silica content of less than 45% by weight.

vein. A mineral deposit in a crevice or a fissure. Vein deposits are generally of hydrothermal origin.

volcanic rock. See *igneous rock*.

Index

Page numbers for illustrations that appear outside the main description of a mineral are shown in *italic numerals*.